Software Engineering Project Management Knowledge Areas

Volume 12:

The Engineering of Software Projects

The Engineering of Software Projects

Software Engineering Knowledge Areas

These twelve volumes support the IEEE *Guide to Software Engineering Body of Knowledge (SWEBOK)* and the IEEE *Computer Society Professional Software Engineering Master Certification* examS.

This is a work-in-progress; each of these volumes is not currently available, but is scheduled to be published in the coming year.

—*Richard Hall Thayer*

Volume 1 — Software Engineering Requirements

Volume 2 — Software Engineering Design

Volume 3 — Software Engineering Construction

Volume 4 — Software Engineering Testing

Volume 5 — Software Engineering Maintenance

Volume 6 — Software Engineering Configuration Management

Volume 7 — Software Engineering Management

Volume 8 — Software Engineering Processes

Volume 9 — Software Engineering Models and Methods

Volume 10 — Software Engineering Quality Assurances

Volume 11 — Software Engineering Economics

Volume 12 — Software Engineering Project Management

These volumes will be published and sold through Amazon Books.

Software Engineering Project Management Knowledge Areas

Volume 12:

The Engineering of Software Projects

Richard Hall Thayer, PhD, CSDP

Contributing authors:

Donald J. Reifer, Reifer Consultants, LLC
Heinz Weihrich, PhD, University of San Francisco

Software Management Training
Carmichael, California
2017

Table of Contents
Software Engineering Project Management

A Partial List of General Abbreviations
(One-of-a-kind abbreviations are identified in place)

a.k.a. — also known as

API — application programming interface

a.s.a.p. — as soon as possible

ConOps — concept of operations (document)

CSCP — Computer Society Certificates of Proficiency

DFD — data flow diagram

FSM — finite state machines

HCI — human computer interface

HW — hardware

I/O — input/output

IDE — integrated development environment

IV&V — independent verification and validation

KA — knowledge area

PSEM — Professional Software Engineering Master (Certification)

SCM — software configuration management

SED — software engineering design

SEM — software engineering management

SEPM — software engineering project management

SER — software engineering requirements

SET — software engineering testing

SQA — software quality assurance

SW — software

SRE — software requirements engineering

SWE — software engineering

SWEBOK — (Guide to the) Software Engineering Body of Knowledge

TBD — to be determined/done

V&V — verification and validation

Honorary Foreword

*To explain the origin of the term "software engineering,"
the following story is offered.[1]*

In the mid-1960s, there was increasing concern in scientific quarters of the Western world that the tempestuous development of computer hardware was not matched by appropriate progress in software development. The software situation looked to be more turbulent. Operating systems had just become the latest rage, but they showed unexpected weaknesses. The uneasiness had been articulated in the NATO Science Committee by its U.S. representative, Dr. I.I. Rabi, the Nobel laureate and famous, as well as influential, physicist. In 1967, the Science Committee set up the Study Group on Computer Science, with members from several countries, to analyze the situation.

The German authorities nominated me for this team. The study group was given the task of "assessing the entire field of computer science," with particular elaboration on the Science Committee's consideration of "organizing a conference and, perhaps later on setting up an "International Institute of Computer Science."

The study group, concentrating its deliberations on actions that would merit an international rather than a national effort, discussed all sorts of promising scientific projects. However, it was rather inconclusive on the relation of these themes to the critical observations mentioned above, which had guided the Science Committee in creating the study group.

Perhaps not all members of the study group had been properly informed about the rationale for its existence. In a sudden mood of anger, I remarked, "The whole trouble comes from the fact that there is so much tinkering with software. It is not made in a clean fabrication process." When I found out that this remark was shocking to some of my scientific colleagues, I elaborated on the idea with the provocative saying, "What we need is *software engineering*."

This remark caused the creation of the expression "software engineering," which seemed to some to be a contradiction in terms, to be stuck in the minds of the members of the group. In the end, in late 1967, the study group recommended that we hold a Working Conference on Software Engineering, and I was made chairman. I not only had the task of organizing the meeting (which was held from October 7 to October 10, 1968, in Garmisch, Germany), but I had to set up a scientific program for a subject that was suddenly defined by my provocative remark.

---

1. Dr. Bauer originally wrote this paper as an introduction to a 1993 IEEE tutorial: *Software Engineering: A European Perspective*, R.H. Thayer, and A.D. McGettrick, eds., IEEE Computer Society Press, Los Alamitos, CA, 1993.

I enjoyed the help of my co-chairmen, L. Bolliet from France and H.J. Helms from Denmark. In addition, I had the invaluable practical support of the program committee members, A.J. Perlis and B. Randell in the section on design, P. Naur and J.N. Buxton in the section on production, and K. Samuelson, B. Galler, and D. Gries in the section on service.

Among the 50 or so participants, E.W. Dijkstra was dominant. Not only did he make cynical remarks like, "The dissemination of error-loaded software is frightening", and, "It is not clear that the people who manufacture software are to be blamed. I think manufacturers deserve better, more understanding users," but he also said, at this early date, "Whether the correctness of a piece of software can be guaranteed or not depends greatly on the structure of the thing made." He had very fittingly named his paper, "Complexity Controlled by Hierarchical Ordering of Function and Variability," introducing a theme that followed his life over the next 20 years. Some of his words have become proverbs in computing, like, "Testing is a very inefficient way of convincing oneself of the correctness of a program."

With the wide distribution of reports from the Garmisch Conference and in a follow-up conference in Rome, from October 27 to October 31, 1969, it happened that not only the term "software engineering" but also the idea behind this term became fashionable. Chairs were created, institutes were established (although the one that the NATO Science Committee had proposed did not come about because of reluctance on the part of Great Britain to have it organized on the European continent), and a great number of conferences were held.

The tutorial nature of the papers in this book is intended to offer readers an easy introduction to the topics and indeed to the attempts that have been made in recent years to provide them with the *tools,* both in a handcraft and an intellectual sense, which allow them now to honestly call themselves *software engineers.*

Friedrich L. Bauer, PhD
Professor Emeritus
Technische Universität München (TUM)
Germany

P.S. In 1989, I met Dr. Friedrich L. Bauer, Professor Emeritus, Universität München, when delivering a software engineering seminar in Munich for the IEEE. Professor Bauer later provided me with the story detailing how he came to name what we now call software engineering. I reprinted the story as an Honorary Foreword by Professor Bauer. Professor Bauer recently passed away (2015) at the age of 90. — RHT

Preface

Software engineering project management is the process of determining what is to be produced in a software system. It encompasses the widely recognized goal of determining the needs for, and the intended external behavior of, a system design. University students as well as candidates for the IEEE Computer Society Certificate of Proficiency exam in *software engineering project management* should focus on the following subareas of the project management knowledge areas (KAs) [www.computer.org/web/education/certifications /2015]:

- Project management is management.

- Planning a software engineering project.

- Organizing a software engineering project.

- Staffing a software engineering project.

- Directing (Leading) a software engineering project.

- Controlling a software engineering project.

To accommodate both groups (university students as well as candidates for the IEEE Computer Society Certificate of Proficiency exam in *software engineering project management*, a software engineering principle that is not included in SWEBOK 2014 and is <u>not</u> likely to produce an exam question is marked with the following statement: "*Note: SWEBOK does not include (to be filled in) in the SWEBOK guide.* The certificate candidate is free to skip this entry. The university student should not.

Regarding references to SWEBOK in this book, when the second edition of the SWEBOK *Guide* is referenced, it is labeled [SWEBOK 2004]; accordingly, when the third edition of the SWEBOK *Guide* is referenced, it is labeled [SWEBOK 2014].

> *A review of both of these books shows that the technical material contained in these two volumes is many times, word for word, identical.*

I need to point out that SWEBOK 2004 is, for all practical purposes, not copyrighted. In contrast, SWEBOK 2014 is copyrighted. Therefore, it is expedient for me to show my references, when identical in both SWEBOKs, as [SWEBOK 2004] as long as I comply with IEEE's usage limitations.

This volume is divided into five chapters followed by a comprehensive index.

1. Chapter 1 is based on the "classic management" knowledge area published by H. Koontz, C. O'Donnell, and more recently, H. Weihrich. This book has repurposed the management functions to reflect the management of a software project.

2. Don Reifer, a well-known project management consultant, wrote Chapter 2, *Principles of Software Engineering Project Management*, a general paper discussing software project management which supports the Koontz, O'Donnell, and Weihrich management theories.

3. Heinz Weihrich has modified Chapter 1 of *Management: A Global Perspective*, 10th ed., edited by Heinz Weihrich and Harold Koontz, for inclusion as Chapter 3 for this book.

4. Chapter 4 is modified *from IEEE Standard for Software Project Management* Plans. This standard has been developed for a software engineering classroom. This classroom standard should not be used to satisfy a commercial software engineering contract. Nevertheless, it can be used as (1) an educational tool and (2) a classroom standard for students to use when preparing classroom software project management specifications.

5. Chapter 5 contains twenty sample exam questions that will guide both certification exam takers and university students.

6. An index of key requirements, terms, documents, tools, authors, and contributors is located at the end of the volume.

Richard Hall Thayer, PhD, CSDP
Life Fellow of the IEEE
Member of the IEEE Computer Society Golden Core
Emeritus Professor of Software Engineering,
   Sacramento State University, California

Acknowledgments

No successful endeavor has ever been undertaken by one person alone. I would like to thank the following people who encouraged and supported me in this effort.

I first want to thank my wife Mildred for her high degree of tolerance as I worked seven days a week on this manuscript. Without her tolerance and support, this book could never have been completed.

I want to thank Ellen Sander for performing copy editing, Jon Digerness of North Coast Graphics for providing me with the comic illustrations, and Jim Tozza for giving me hardware and software support.

In addition, I want to thank Steve Tockey for providing me with numerous tips about the Computer Society exam specifications in order to maximize the usefulness of our software engineering textbook and SWE guidebook, and Melville (Mel) Piercey of Copy Plus for providing cover artwork and designing and drawing the engineering chapter graphics.

Finally, I want to thank our little dog Maxwell (a.k.a. Max, Maxie, Maxcito, or Speedy) who kept me company in the evening hours when everybody else had gone to bed.

A happy Max says that:

This is a Terrrrrrrific Book. I chewed on a copy,
and it was tasty.

A Note to Our Readers

One of the advantages of using a "print-on-demand" (POD) publishing service is the ability to make manuscript changes relatively easily when errors or improvements are identified.

The authors encourage you to identify and send potential errors or suggested improvements to the e-mail address listed below. Although I do not guarantee to make all the changes identified, I do promise to review and seriously consider all recommendations.

Disclaimer

While I have more than 50 years of software engineering experience, including university teaching, I am not a technical expert in every component of software engineering. To make up for this shortcoming, I have made extensive use of material written by subject matter experts and papers (many posted on the web) as source documents.

Every effort has been made to make this software engineering reference as complete and accurate as possible. However, I can make no representation or warranties with respect to accuracy or completeness of the contents of this book and specifically disclaim any implied warrantee of merchantability or fitness for a particular purpose. The advice and strategies contained herein may not be suitable for your situation. If in doubt, you should consult with a professional software engineer. Where appropriate, neither I nor the printer will be liable for the loss of profit or other commercial damages, including but not limited to, special, incidental, consequential, or other damages [IEEE Press disclaimer].

Please keep me posted.

Richard Hall Thayer, PhD, CSDP
Life Fellow of the IEEE
Member of the IEEE Computer Society Golden Core
Emeritus Professor of Software Engineering,
  Sacramento State University, California

thayer@csus.edu

Chapter 1
Software Engineering Project Management Fundamentals

This chapter is a textbook and study guide introducing the principles and common problems associated with software engineering project management (SEPM).

This volume can be used either to supplement a university course in software engineering project management (SEPM) or as a study guide to aid individual software engineers studying for the IEEE Professional Software Engineering Master (PSEM) Certification exams in software management.

Management involves the activities and tasks undertaken by one or more persons for the purpose of planning the activities of others in order to achieve objectives that could not be achieved by the others acting alone. This volume describes management, the universality of management, and the functions, activities, and tasks of conducting a SWE management effort.

This volume places emphases on software engineering project management. However, most of the management functions, processes, activities, and management issues can generally be applied to all elements of software engineering management.

INTRODUCTION

Project management (PM) is a system of management procedures, practices, technologies, skills, and experiences that are necessary to successfully manage an engineering project. The act of managing a software project is called software engineering project management (SEPM). These managers can be called "software project managers," or simply "project managers."

Project management can be defined as the application of management fundamentals—*planning, organizing, staffing, directing,* and *controlling*—to ensure that software adheres to user specifications, development, and maintenance.

It should be noted "early on" that I do not agree with the SWEBOK or the PMBOK descriptions of the Software Engineering Project Management Knowledge Area *(KA).* The SWEBOK description of project management is not wrong; it is just incomplete. This issue was not as important in early editions of this book that were being used only as study guidebooks for the CSDP qualifying exam. However, these earlier editions could not be used as textbooks for a university undergraduate or graduate level course in software engineering project management.

The premise of this book is that managing a SWE project is the same as managing any other endeavor or organization.

This volume contains the following six SEPM knowledge area (KA) subareas:

(1) **Project management** is the *planning, organizing, staffing, directing,* and *controlling* of software project activities.

(2) **Planning** is predetermining a course of action for accomplishing organizational objectives.

(3) **Organizing** is arranging the relationships among work units for accomplishing objectives, and granting responsibility and authority to individuals to attain those objectives.

(4) **Staffing** is selecting and training people for positions in the organization.

(5) **Directing** (a.k.a. *leading*) is creating an atmosphere that will assist and motivate people to achieve desired results.

(6) **Controlling** is establishing, measuring, and evaluating performance of project activities to achieve planned objectives.

SOFTWARE ENGINEERING PROJECT MANAGEMENT

This volume is based on my belief that the management of a SWE project can be accomplished by implementing management process similar to skills used by other enterprises or organizations. Figure 1 provides a top-level decomposition and breakdown of the SEPM KAs. Figure 2 shows a sequence of software development phases, their relationships with each other, and the major software product resulting from each phase. Planning is generally developed during the early part of the requirements engineering phase. The phases and products associated with SEPM are marked with a "star."

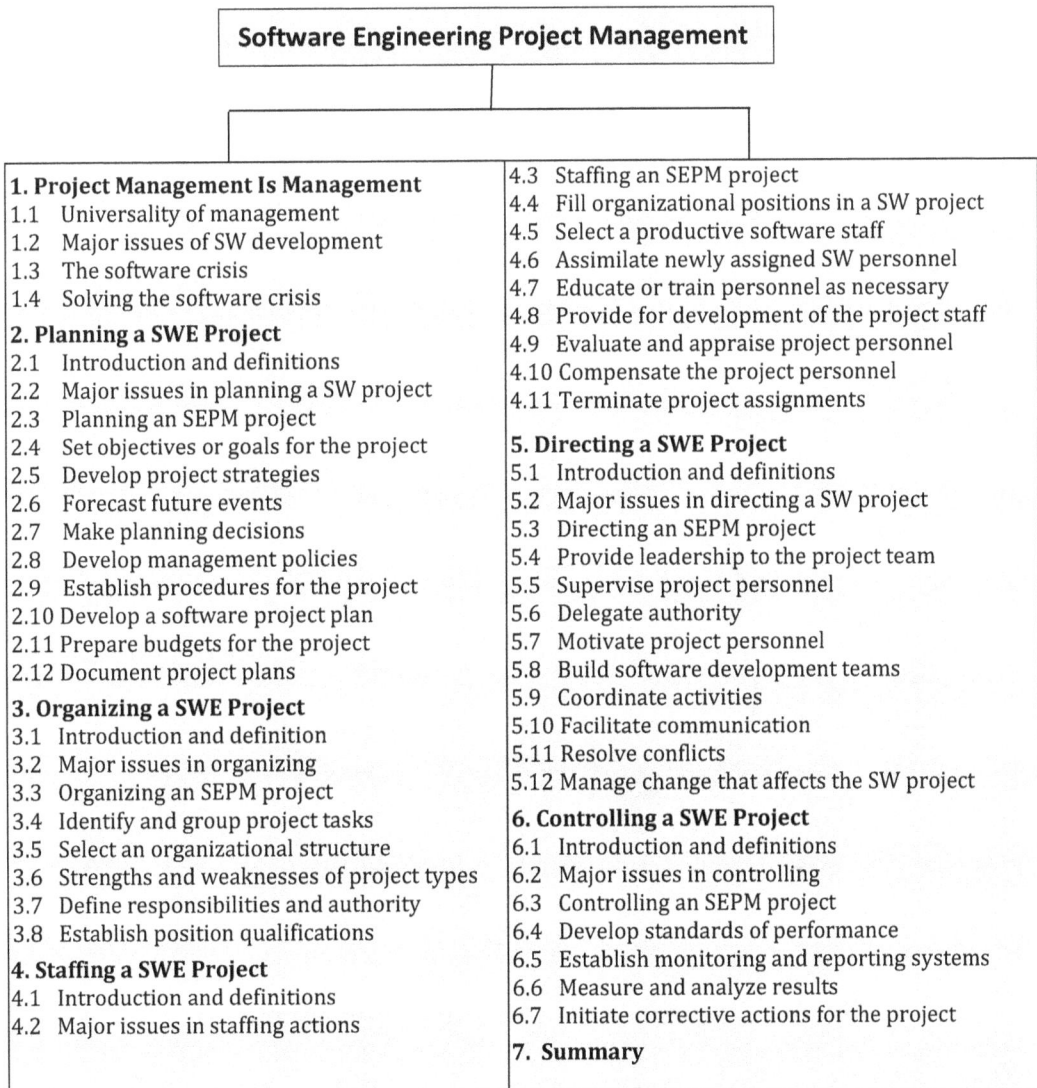

```
Software Engineering Project Management
```

1. Project Management Is Management
1.1 Universality of management
1.2 Major issues of SW development
1.3 The software crisis
1.4 Solving the software crisis

2. Planning a SWE Project
2.1 Introduction and definitions
2.2 Major issues in planning a SW project
2.3 Planning an SEPM project
2.4 Set objectives or goals for the project
2.5 Develop project strategies
2.6 Forecast future events
2.7 Make planning decisions
2.8 Develop management policies
2.9 Establish procedures for the project
2.10 Develop a software project plan
2.11 Prepare budgets for the project
2.12 Document project plans

3. Organizing a SWE Project
3.1 Introduction and definition
3.2 Major issues in organizing
3.3 Organizing an SEPM project
3.4 Identify and group project tasks
3.5 Select an organizational structure
3.6 Strengths and weaknesses of project types
3.7 Define responsibilities and authority
3.8 Establish position qualifications

4. Staffing a SWE Project
4.1 Introduction and definitions
4.2 Major issues in staffing actions

4.3 Staffing an SEPM project
4.4 Fill organizational positions in a SW project
4.5 Select a productive software staff
4.6 Assimilate newly assigned SW personnel
4.7 Educate or train personnel as necessary
4.8 Provide for development of the project staff
4.9 Evaluate and appraise project personnel
4.10 Compensate the project personnel
4.11 Terminate project assignments

5. Directing a SWE Project
5.1 Introduction and definitions
5.2 Major issues in directing a SW project
5.3 Directing an SEPM project
5.4 Provide leadership to the project team
5.5 Supervise project personnel
5.6 Delegate authority
5.7 Motivate project personnel
5.8 Build software development teams
5.9 Coordinate activities
5.10 Facilitate communication
5.11 Resolve conflicts
5.12 Manage change that affects the SW project

6. Controlling a SWE Project
6.1 Introduction and definitions
6.2 Major issues in controlling
6.3 Controlling an SEPM project
6.4 Develop standards of performance
6.5 Establish monitoring and reporting systems
6.6 Measure and analyze results
6.7 Initiate corrective actions for the project

7. Summary

Figure 1: Hierarchical listing of topics for the SWE project management KA

1. Project Management Is Management.

SWE projects are frequently part of larger, more comprehensive projects that include equipment (hardware), facilities, personnel, and procedures, as well as software. Examples include projects to develop aircraft systems, accounting systems, radar systems, inventory control systems, and railroad switching systems. Some systems engineering projects are typically managed by one or more system project managers (sometimes called "program managers") who manage projects composed of engineers, domain experts, scientific specialists, programmers, support personnel, and others. If the software to be delivered is a "stand-alone" software system (a system that does not involve development of other, nonsoftware components), then the SWE project manager may be called a "system project manager."

1.1 Universality of management.

The *universality of management*, a concept that originated with management science [Fayol 1949] [Koontz & O'Donnell 1972], means we can apply all references to the application of general management functions "across-the-board". The universality of these concepts provides a management framework for adapting traditional management functions to project management [Thayer & Pyster 1984]. It is from these original general management functions that this chapter derives the detailed activities and tasks that should be undertaken by a manager assigned to an SWE project.

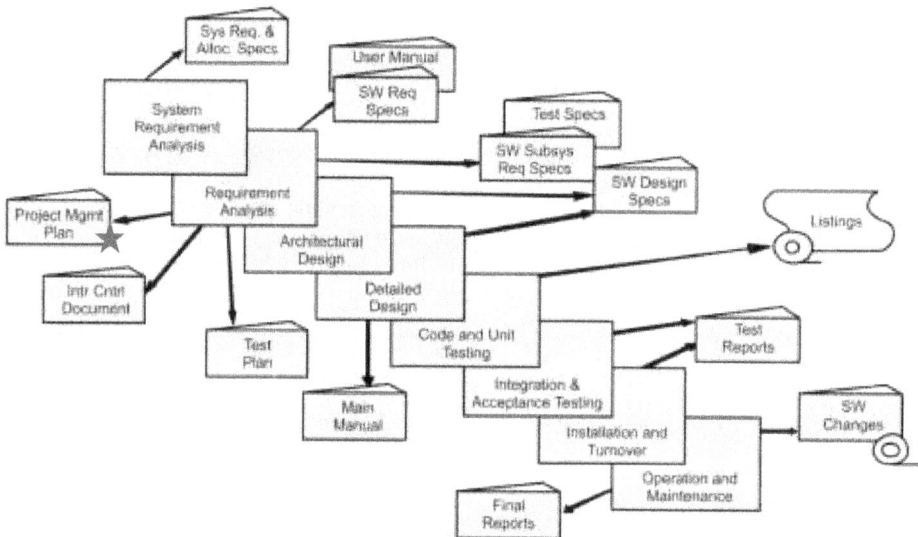

Figure 2: Sequence of software development phases

Figure 3 depicts the classic management model as portrayed by such well-known authors in the field of management as Koontz and O'Donnell [1972]. Even though Koontz and O'Donnell represent the classic references of management science, other authors have also contributed to this area of knowledge, such as Fayol [1949], MacKenzie [1969], Cleland and King [1972], Donnelly, Gibson and Ivance-vich [1975], Koontz, O'Donnell and Weihrich [1980], Rue and Byars [1983], Thayer and Pyster [1984], Thayer [1988, 2000], Weihrich [1993] and Reifer [2006].

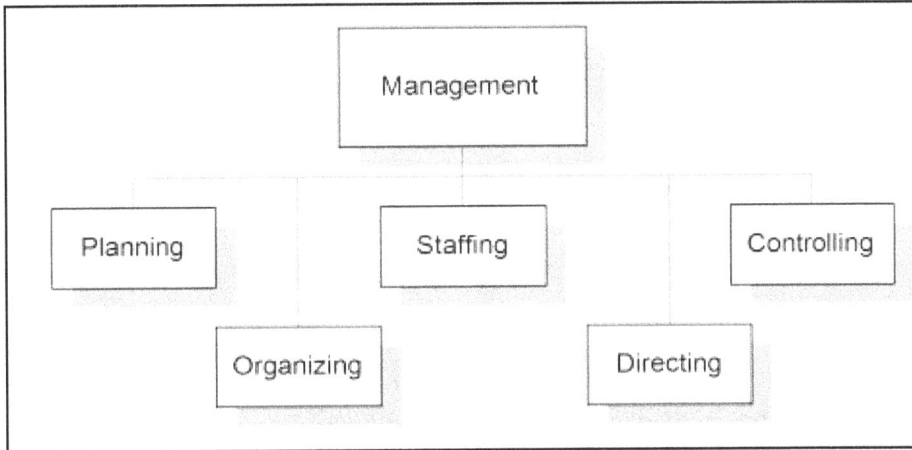

Figure 3: The classic management model

According to our SEPM model, management is partitioned into five separate functions or components: *planning, organizing, staffing, directing,* and *controlling*. Table 1 lists and defines these functions.

Table 1: The classic management model defined

Functions	Definition or Explanation
Planning	Predetermining a course of action for accomplishing organizational objectives.
Organizing	Arranging the relationships among work units for accomplishment of objectives and granting responsibility and authority to attain those objectives.
Staffing	Selecting and training personnel to staff the organization.
Directing (Leading)	Creating an atmosphere that will assist and motivate personnel and staff to achieve desired results.
Controlling	Establishing, measuring, and evaluating performance of activities toward planned objectives.

One of the most compact descriptions of management was developed by Harvard's MacKenzie [1969].

In addition, a search of the World Wide Web using "planning," "organizing," "staffing," "directing," and "controlling" illustrates that numerous papers, articles, and courses recognize project management as being composed of these five functions. In addition, these five management functions or components can be further divided into tasks, which are still more detailed.

This management approach applies to any organization and to managers at all organizational levels. Management principles apply to both small and large organizations, to profit and not-for-profit enterprises, and to the manufacturing and service industries.

An analysis of such diverse organizations as a McDonald's restaurant, the San Francisco 49ers, the State of California, a U.S. Air Force fighter squadron, a software engineering project, the country of Brazil, and a family unit of man, woman, four children, and a dog will each identify with these same five functions—only the tasks, tools, and approach would be different.

My professional position is that management functions should be applied to any type of software project. Use of this application approach is not restricted by the size, complexity, or criticality of the software product.

1.2 Major issues of software development.

The major issues in software development involve the failure of software engineers to use a rigorous software development process. Failure to incorporate a software development process is because either the company does not have standard practices in place or, if it does, they are not being followed. Managers frequently manage by implementing processes and practices that were apparently successful during their last project [Paulk et al. 1996].

1.3 The software crisis.

The "software crisis" was a term used in the early days of software development to highlight the difficulty of writing useful and correct computer programs within required time and cost schedules. The software crisis is generally credited to the rapid increases in computer capability and the complexity of the problems that could be undertaken. With the increase in computing power, many software problems arose because existing methods were neither sufficient nor effective.

At the 1968 NATO Software Engineering Conference in Garmisch-Partenkirchen, Germany, Edsger Dijkstra's 1972 ACM Turing Award Lecture referred to this problem:

> *The major cause [of the software crisis] is that the machines have become several orders of magnitude more powerful! To put it quite bluntly: as long as there were no machines, programming was no problem at all; when we had a few weak computers, programming became a mild problem and now that we have gigantic computers, programming has become an equally gigantic problem [Dijkstra 1972].*

A Report to the Defense Science Board Task Force on Military Software states that today's "major problems with software development are not technical problems, but management problems" [Brooks 1987].

The causes of the software crisis were linked to the overall complexity of the software development process. The crisis manifested itself in several ways [https://www.en.wikipedia.org/wiki/Software_crisis]:

- Projects were exceeding their budget.
- Projects were running late.
- Software often did not meet requirements.
- Projects were unmanageable and code was difficult to maintain.
- The software was never delivered.

The State of California, for one, has certainly had its share of software problems. In a recent article published in the *Sacramento Bee*, the State Auditor reported that California's state technology departments lack guidelines for managing and/or terminating failing software engineering projects.

Still, from 1994 to 2013, California State government spent $985 million on seven computer projects that either were terminated or suspended [Sacramento Bee, 2015b]. For example:

(1) A $40M Department of Motor Vehicles project attempted to merge existing, separate systems to manage driver's licenses and vehicle registrations. The project was never delivered. A state senator was so disgusted with the results of the project that he suggested building a bonfire with the partially delivered hardware [Neumann 1993].

(2) A Correctional Management Information System was initiated to manage the state's prisoners and parolees, at a cost of $44M. The V&V contactor said it could not be built from the design document. The State sued the software development company for $25M and won [Sacramento Bee, 1997b].

(3) A State Automated Child Support System was developed at a cost of $100M. After eight years, the project was canceled [Sacramento Bee, 1997a.] The state paid $1B in federal penalties because of delays in implementing this system [Sacramento Bee, 2015a].

(4) The State of California has attempted an overhaul of its 30-year-old Unemployment Insurance payment processing system—part of a $188 million project that eventually would allow claimants to submit required certifications for jobless benefits online or by telephone. The project requirements involved converting the manual check payment system to a new automated system.

Converting the backlog of claims took much longer than anticipated and nearly 150,000 Californians—about 19% of jobless benefit recipients—had their unemployment checks delayed. The numbers were so significant that they skewed the U.S. Labor Department's unemployment reports [Sacramento Bee, 2013a].

However, the news is not all bad. The Standish Group's 2012 CHAOS Manifesto results show an increase in project success rates, with 39% of all projects succeeding (delivered on time, within budget with required features, and functions). However, 43% of projects were challenged (for being delivered late, over budget and/or with less than the required features), and 18% failed, meaning they were cancelled prior to completion or delivered and never used) [http://www.versionone .com/assets/img/files/ChaosManifesto2013 .pdf].

A 39% increased success rate is an improvement, but is it good enough?

These problems together have been called the "software crisis," which is characterized by software that is delivered *late, over budget, and fails to meet the customer's system requirements* [Gibbs 1994]. Many, if not most of these problems, have been blamed on inadequate SEPM.

> *A major source of software development problems can be attributed to poor or nonexistent software engineering project management. Each chapter of this book addresses methods to successfully manage a software engineering project.*

1.4 Solving the software crisis.

Software development requires a well-defined process to develop a software system—i.e., software engineering and software engineering project management. Established SWE techniques such as in-depth requirements analysis, inspections, reviews, testing, and documentation are reduced or eliminated when the project falls behind in cost or schedule and/or the customer demands more functionality without an increase in budget or schedule [Paulk et al. 1996].

A Department of Defense (DoD) Software Initiative describes the DoD's motivation for establishing the software initiative and reviews its goals and objectives. The plan is a combination of technology transfers, incentive programs to build software expertise, and various quality control measures, which are expected to help alleviate some of the DoD's software ills [DoD Software Initiative 1983].

The ultimate software crisis is a "software project failure." A software project is considered a "failure" if *any one* of the following is true:

(1) The project is late.

(2) The project is over budget.

(3) The project does not meet its specified requirements.

Note that the software crisis is project failure, not software failure.

2. Planning a Software Engineering Project

Planning is deciding in advance what to do, how to do it, when to do it, and who is to do it [Koontz and O'Donnell 1972].

2.1 Introduction and definitions.

Planning a SWE project consists of the management activities that lead to selection, among alternatives, of future courses of action for the project and a program for completing those actions. Planning thus involves specifying the goals and objectives for a project and the strategies, policies, plans, and procedures required for achieving them. Every SWE project should start with an effective and complete plan. Uncertainties and unknowns, both within the software project environment and from external sources, make planning necessary. Planning focuses attention on project goals, actions necessary to reach those goals, and potential risks and problems that might interfere with obtaining those goals.

2.2 Major issues in planning a software project.

Some of the major issues when planning a SWE project are:

(1) Software requirements are frequently incorrect and incomplete; therefore, it becomes difficult to assign efforts to complete the required tasks.

(2) Many software requirements specifications are unstable and are subject to frequent and major changes.

(3) Planning is often not attempted in the mistaken belief that it is a waste of time because the plans will change anyway.

(4) Planning for schedule and cost are not based on project requirements or development environment.

(5) It is difficult to estimate the size and complexities of the software project in order to make a realistic cost and schedule estimate.

(6) Cost and schedule are not re-estimated when changes are made to requirements or to the development environment.

(7) Risk factors are not assessed, documented, or managed.

(8) Most software development organizations do not collect project data on past projects (for use in future project estimates).

(9) Companies do not establish software development policies or processes.

2.3 Planning an SEPM project.

Planning is the selection from among alternatives of future courses of action for the enterprise as a whole and within each department [Koontz and O'Donnell 1972]. Planning involves selecting missions and objectives and the actions required to achieve them; it requires decision making, which is choosing future courses of action from among alternatives. There are various types of plans, ranging from overall purposes and objectives to the most detailed actions to be taken, such as to order a special stainless steel bolt for an instrument or to train workers for an assembly line.

Table 2 provides an outline of the planning activities that must be accomplished by software project managers when planning their projects. The project manager is responsible for developing numerous types of plans.

Table 2: Planning activities for software engineering projects

Activity	Definition or Explanation
Set objectives or goals	Determine the desired outcome for the project.
Determine development strategies	Develop, analyze, and/or evaluate different ways to meet project objectives and goals.
Forecast future events	Anticipate possible future events such as unexpected problems and/or potential risks. Determine the impact of these events on possible development plans and strategies.
Make planning decisions	Select a course of action from among alternatives. Build contingency plans.
Develop policies	Make standing decisions on important recurring matters to provide a guide for decision-making.
Develop procedures	Detail the manner in which a project activity must be accomplished.
Develop project plans	Establish policies, procedures, tasks, schedules, and resources necessary to complete the project.
Prepare budgets	Allocate estimated costs to project functions, activities, and tasks.
Document project plans	Record policy decisions, courses of action, budget, program plans, and contingency plans.

2.4 Set objectives or goals for the project.

Objectives or *goals* are the ends toward which an activity is aimed [Koontz and O'Donnell 1972]. The first planning step for a SWE project is to determine what the project must accomplish, when it must be accomplished, and what resources are necessary.

The primary objective of a SW project is (for the third time) to be able to finish and deliver a software system to a customer:

- On or before the system delivery date promised in the contract.

- Within the cost or labor hours promised or agreed to by contract.

- Achieving delivery of the promised or contractually agreed system requirement.

2.5 Develop project strategies.

Another software planning activity is to develop and document a set of management strategies to develop the system for which the project exists. The project manager must establish the necessary conditions for the project to be a success. *(See Table 3 for examples of the many different types of plans.)*

Table 3: Types of plans for software projects

Type of Plan	Definition or Explanation
Objective	The project goals toward which the activities are directed.
Strategic	The overall approach to a project that provides guidance for placing emphases and using resources to achieve the project objectives.
Policy	Policies are established by management to guide in decision-making and project activities. Policies limit the freedom of making decisions but allow for some discretion.
Procedural	Directives that specify customary methods of handling activities; guides to actions rather than decision-making. Procedures detail the exact manner in which a project activity must be accomplished. They allow for very little discretion.
Rule	Requirements for specific and definite actions to be taken or not taken with respect to particular project situations. No discretion is allowed.
Project plan	An interrelated set of goals, objectives, policies, procedures, rules, work assignments, resources to be used, and other elements necessary to manage a software project.
Budget	A statement of constraints for resources expressed in quantitative terms such as dollars or staff-hours.

2.6 Forecast future events.

Determining future courses of action will be based on the project status and environment as well as the project manager's vision of the future [Koontz and O'Donnell 1972].

Determining the future, called "forecasting," can be addressed in two steps:

Step 1 involves predicting future events, such as risk analysis, availability of personnel, unexpected problems, the availability of new computer hardware, and perhaps potential project risks, followed by (Step 2) predicting the impact of these future events on the project.

2.7 Make planning decisions.

The project manager will conduct a feasibility analysis to develop a clear description of project objectives and evaluate alternative approaches in order to determine whether the proposed project is the best solution to the problem given the constraints of technology, resources, finances and social/political considerations [SWEBOK 2004]. In most projects, there is more than one way to conduct the project—but not with equal cost, equal schedule or equal risk. It is the project manager's responsibility to examine ways to meet the project objectives.

For example, one approach might be very costly in terms of personnel and machines yet may reduce the schedule dramatically. Another approach might reduce both schedule and cost but severely risk not being able to deliver a satisfactory system. A third approach might be to stretch the schedule, thereby reducing the cost of the project.

The manager must examine each course of action to determine advantages, disadvantages, risks, and benefits.

The project manager, in consultation with higher-level management, the customer, and other appropriate parties, is responsible for selecting the best course of action for meeting project goals and objectives. The project manager is responsible for making tradeoff decisions involving cost, schedule, design strategies, and risks.

For example, in *Agile projects,* schedule and resources are considered fixed but features are prioritized—if necessary, lower-priority features are postponed to later developments. Having thus determined what features will be developed in the next increment; a detailed plan is developed for that increment [SWEBOK 2004].

The project manager is responsible for selecting the best development model for the project. For example, possible models might include the waterfall model, the incremental or evolutionary model, and Boehm's spiral model. These models recognize that it may be impossible to estimate accurately, at the beginning of a project, the resources and schedule required to meet the project's functional and nonfunctional requirements.

Therefore, the team members regularly (usually at specified intervals) review progress to date. If initial estimates of schedule and resources are not being met, the project manager needs to make appropriate modifications to plans.

The project manager is responsible for allocating equipment, facilities, and people associated with the scheduled tasks, including the allocation of responsibilities for completion. This activity is both informed and constrained by the availability of resources and their optimal use under these circumstances, as well as by issues relating to personnel (for example, productivity of individuals/teams, team dynamics, organizational and team structures) [SWEBOK 2004].

The project manager is responsible for establishing quality processes for the project. Quality is defined in terms of pertinent attributes of the specific project and any associated products, in either quantitative or qualitative terms.

These quality characteristics will have been determined in the specification of detailed software requirements. Thresholds for adherence to quality are set for each indicator as appropriate to meet stakeholder expectations for the software at hand. Procedures relating to ongoing software quality assurance (SQA) throughout the process and for product (deliverables) verification and validation are also specified at this stage. Quality thresholds, which are set for each indicator by stakeholder expectations, must be adhered to as appropriate.

The project manager is responsible for the following four quality processes:

(1) The project manager is responsible for establishing management of the project plan—how the project will be managed and how the plan overseeing the project will be managed. Reporting, monitoring, and control of the project must fit the selected SWE process and the expectations of the project. They must also reflect the various artifacts that will be used for managing the project. Nevertheless, in an environment where change is an expectation rather than a shock, the management of plans is also vital.

Plans need to be systematically used, monitored, reviewed, reported, and where appropriate, revised. Plans associated with other management-oriented support processes (for example, documentation, software configuration management, and problem resolution) also need to be appropriately managed [SWEBOK 2004].

(2) The project manager is responsible for selecting the methods and tools, both technical and managerial, by which the project will be managed, and the product developed. For example, will the requirements be documented using structured analysis methods or use cases? Will testing be done top-down, bottom-up, or both? Which tools, techniques, and procedures will be used in planning the schedule: PERT, CPM, workload chart, work breakdown chart (WBS), or Gantt chart?

The products of each task (for example, architectural design, and inspection report) are specified and characterized. To use software components from previous developments or to utilize off-the-shelf software products are evaluated. Use of third parties and procured software are planned and suppliers are selected by the project manager [SWEBOK 2004].

(3)   The project manager is responsible for developing or procuring process standards (in contrast to product standards) that can be used to establish procedures. Process standards may be adopted from corporate standards or written for a particular project. Process standards might cover topics such as reporting methods, reviews, and documentation preparation requirements [IEEE Software Engineering Standards].

(4)   The project manager is responsible for establishing a project costs and schedule beginning with the project initiation and ending with a completion or delivery date. The project activities are undertaken according to the schedule. Resources are utilized (for example, personnel, effort, and funding) and deliverables are produced (for example, architectural design documents and test cases).

2.8 Develop management policies.

Policies are predetermined management decisions [Koontz and O'Donnell 1972]. The project manager may establish policies for the project to provide guidance to supervisors and individual team members making routine decisions. For example, it might be a policy of the project that status reports from team leaders are due in the project manager's office by close of business each Thursday.

Policies can reduce the need for interaction pertaining to every decision and provide a sense of direction for the team members. In many cases, the project manager does not develop new policies for the project but follows some or all of the policies established at the corporate level.

As an early example, the management of a well-known and respected system engineering company—TRW—discovered that they did not have an "in-house" policy for developing *software systems*. This was particularly troublesome because TRW was famous for taking other software developers "to task" for poor software development procedures.

To make amends, senior TRW managers implemented an SEID Software Development Policy. (SEID was the acronym for the TRW software development organization.) The policy reads in part:

> *SEID Software Development Policies shall be maintained by the SEID Software Policy Change Control Board. The chair of this Board shall be appointed by the SEID General Manager. Board members shall be appointed in writing by each Operation and Business Area Manager. Proposed changes to SEID software development policies must be submitted in writing to the Board with the approval of the appropriate Operations or Business Area representative.*

> *At least once each year, the Board shall convene to review the policies in their totality for relevancy and currency. Where indicated, they shall*

propose revisions to the policies, subject to the review and approval of the SEID General Manager [TRW 1977].

See Appendix B for a list of the 1975 TRW software engineering processes.

2.9 Establish procedures for the project.

Procedures for an SWE project are tactics that establish required methods of handling future activities. They are guided to action rather than intelligent thought. In addition, they detail the exact manner in which a certain activity must be accomplished. They are a chronological sequence of required actions [Koontz, O'Donnell & Weihrich 1984].

2.10 Develop a software project plan.

A *project plan* is a complex set of goals, policies, procedures, rules, task assignments, steps to be taken, resources to be employed, and other elements necessary to carry out a given course of action that is supported by necessary capital and delivered within budget [Koontz & O'Donnell 1972]. Typically, the project plan specifies the following:

(1) The procedures and rules for the project that are to be performed by the software development staff in order to deliver the final and correct software product. This usually requires the partitioning of the project activities into small, well-specified tasks. A useful tool for representing the partitioned project is the work breakdown structure (WBS).

(2) How the project will be managed and how the plan will be managed must be specified. Project reporting, monitoring, and controlling must fit the selected SWE process and the practical aspects of the project. In addition, the project plan must reflect the various artifacts that will be used for managing the project. Nevertheless, in an environment where change is an expectation rather than a shock, it is vital that plans are themselves well managed.

Effective project management requires adherence to project plans by systematically directed, monitored, reviewed, reported, and where appropriate, revised project planning. Plans associated with other management-oriented support processes (for example, documentation, software configuration management, and problem resolution) also need to be managed in the same manner [SWEBOK 2004].

The chief systems engineer (or chief designer) assists the project manager by determining detailed project milestones and provides input to develop the schedule.

2.11 Prepare budgets for the project.

Budgeting is the process of placing cost figures in the project plan [Koontz & O'Donnell 1972]. The project manager is responsible for determining the cost of the project and calculating budget allocations to project tasks. Cost is the common denominator for all elements of the project.

Based on the breakdown of tasks and inputs and outputs, the expected effort range required for each task is determined using a calibrated estimation model based on historical size-effort data when available, and relevant, or other methods like expert judgment. Task dependencies are established and potential bottlenecks are identified using suitable methods (e.g., critical path analysis).

Bottlenecks are resolved where possible and the expected schedule of tasks with projected start times, durations, and end times are produced (e.g., PERT chart). Resource requirements (people and tools) are translated into cost estimates. This is a highly iterative activity, which must be negotiated and revised until consensus is reached among affected stakeholders (primarily engineers and management).

Requirements for personnel, computers, travel, office space, and equipment can be compared and cost-tradeoffs made only when these requirements are measured in terms of their monetary value. . This requires that adherence to plans be systematically managed.

2.12 Document project plans.

The project manager is responsible for *documenting* the project plan. He (or she) might also be responsible for preparing other plans such as software quality-assurance plans, software configuration management plans, staffing plans, test plans, and so forth

How the project and the plan will be managed must be documented. Reporting, monitoring, and controlling the project must fit the selected SWE process and the realities of the project and must reflect the various artifacts that are used for managing it. Plans associated with other management-oriented support processes (for example, documentation, software configuration management, and problem resolution) also need to be managed in the same manner (SWEBOK 2004).

The project plan is the primary means of communicating with other entities that interface with the project.

3. Organizing a Software Engineering Project.

The purpose of an organization is to "focus the efforts of many on a common goal" [Donnelly, Gibson & Ivancevich 1975].

3.1 Introduction and definitions.

Organizing an SWE project involves having an effective and efficient organizational structure for assigning and completing project tasks and establishing the authority and responsibility of relationships among the tasks [Koontz & O'Donnell 1972].

Organizing involves itemizing the project activities required to achieve the objectives of the project and arranging these activities into areas of mutual support. It also involves the assignment of these activities to various organizational entities and the delegation of responsibility and authority needed to carry out the activities. *Note: SWEBOK does not include organizing as an SWE management technique.*

It is difficult to determine the best organizational structure for both a project and for the organization conducting the project. According to [Youker 1977], there exists a spectrum of organizational techniques for software projects, ranging from the functional format to the matrix format to the project format. A paper authored by Youker reflects some of the commonalities between functional, matrix, and project structures (*see Figure 4*).

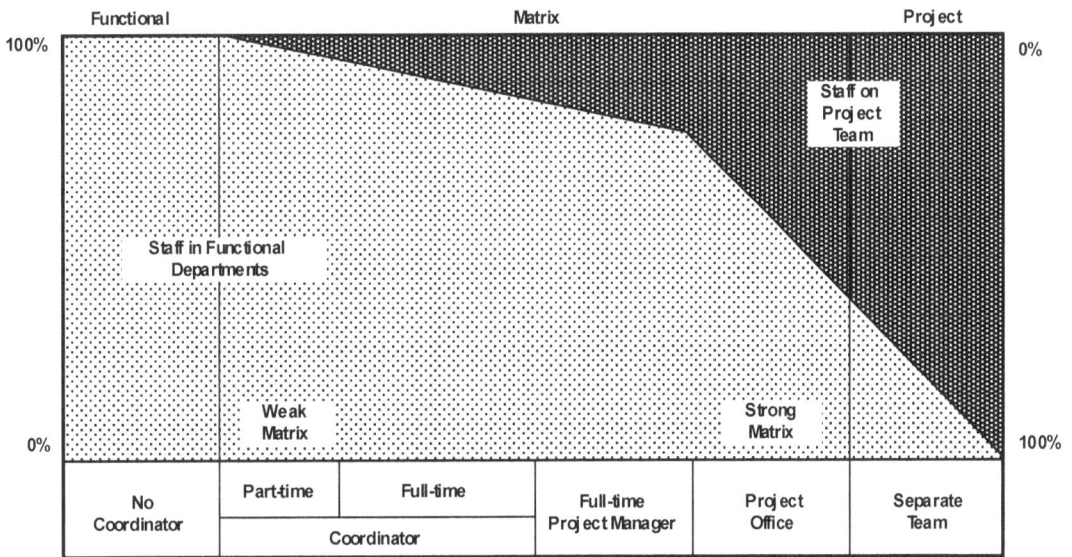

Figure 4: A comparison between functional, matrix,
and project organizations [Youker 1977]

3.2 Major issues in organizing.

The following are major issues encountered when organizing a SWE project:

(1) It is sometimes very difficult to determine the best way to organize a particular project and/or environment.

(2) The organizational structure may be preassigned by upper management with little regard for the job to be done.

(3) An organizational structure may leave responsibilities for some project activities and tasks undefined or unclear.

(4) A matrix organizational structure is not accepted by many SWE managers and software development personnel.

(5) Many team leaders are expected to perform technical tasks (e.g., coding) as well as to manage their team.

3.3 Organizing an SEPM project.

Table 4 provides an outline of the activities that must be accomplished by the project manager responsible for organizing a project [MacKenzie 1969]. The remainder of this section provides detail concerning the activities outlined in Table 4. The selected organizational structure should match the needs and goals of the project with an environment that facilitates communication both internal and external to the organization.

Table 4: Organizing activities for software projects

Activity	Definition or Explanation
Identify and group project functions, activities, and tasks.	Define, size, and categorize the project work.
Select organizational structures.	Select appropriate structures to accomplish the project and to monitor, control, communicate, and coordinate the project.
Create organizational positions.	Establish title, job descriptions, and job relationships for each project role.
Define responsibilities and authority.	Define responsibilities for each organizational position and the authority to be granted for fulfillment of those responsibilities.
Establish position qualifications.	Define qualifications for persons to fill each position.

3.4 Identify and group project tasks.

The manager is responsible for reviewing the project requirements, defining the various tasks to be accomplished, sizing, and grouping those tasks into logical entities. Titles and organizational entities are assigned to the assembly of tasks: for example, analysis, design, coding, and testing. This information enables the project manager to select an organizational structure to control these groups. *See Table 5 for an example of task identification and grouping.*

The project manager must also identify the supporting tasks needed, both internal and external to the project.

Table 5: An Example of task identification and grouping

Project Tasks	Organizational Entity
Determine software system requirements. Partition and allocate software requirements to software components. Develop software architectural design. Identify and schedule tasks to be done.	Software System Engineering
Analyze software components for Product 1. Design components of Product 1. Implement Product 1 software. Prepare documents. Support verification and validation (V&V).	SWE Application Group 1
Execute the same tasks and activities as defined in Product 1 for Product 2.	SWE Application Group 2
Prepare software V&V plans. Conduct V&V activities. Prepare and support software testing.	Software V&V
Establish software quality -assurance plans. Conduct software quality activities. Document results of quality activities.	Software Quality Assurance (SQA)

Examples of internal tasks are secretarial support, word processing support, financial monitoring, and project administration. External to the project, there may be tasks associated with travel requirements, motor pools, security guards, etc.

3.5 Select an organizational structure for the project.

After identifying and grouping project tasks, the project manager must select one of several different organizational structures. (*See Figures 6 through 8.*)

The project manager may not have the luxury of selecting the best project organizational type since this may be determined by policy decided at the corporate level. Regardless of who does the actual work, it is still a project management function. The organizational type that is most approximate to the project and the environment must be selected.

(1) There are two types of conventional organizational structures: *line* and *staff organizations.*

    a. ***Line organizations*** — A *line organization* has the responsibility and authority to perform the work that represents the primary mission of the larger organizational unit. (Figure 5 portrays a line organization provided the general manager reports to a higher-level engineering organization.) Most software projects strive to operate as a line organization. By adopting a line organization, the project is probably more important to the company and less likely to be cancelled due to political or financial problems.

b. **Staff organizations** — In contrast, a *staff organization* is usually composed of a group of functional experts who have responsibility and authority to perform special activities that help the line organization to accomplish its work. All organizations in a company are either line or staff structures. In Figure 5, the groups of system engineering, software engineering teams, and verification and validation are normally line functions. Note that a line organization can work for a staff function and vice-versa.

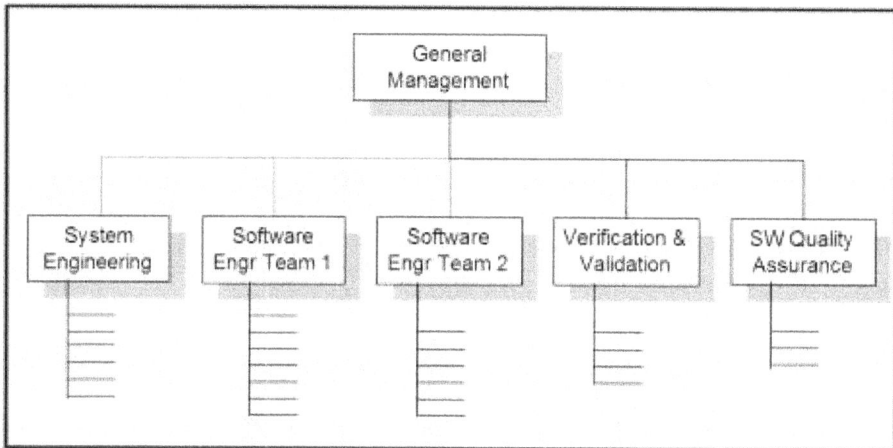

Figure 5: A software line organization

For example, Sacramento State University has a Computer Science Department. This department has a computer facility available for faculty and student use. The university also has a Computer Services Department for the use of the university administration. Since the main activity of the university is academic, the Computer Science Department's computer facility is part of a "line" organization and the university Computer Services Department facility is a "staff" organization.

(2) There are three types of software development project organizational structures: *functional, project,* and *matrix:*

a. **Functional organizational structure** — *Functional project organizational structures* are typically used to develop small to medium sized system or software engineering projects. In Figure 6, the second-level structures portrayed are *functions not organizations. These functions may be assigned singularly or as a group to one or more functional managers. These functional managers typically have the authority to hire, supervise, train, promote, and terminate engineers within their organization.*

The project to be constructed is initially usually assigned to the system functional organization. The project is accomplished by passing the partially completed project from SWE group to SWE group until it is finished. Each functional manager is responsible for completing his "piece of the pie" with his or her resources provided.

Problems erupt when one group believes the previous group did not complete its job properly or adequately. Sometimes the SW project is mistakenly assigned to, i.e., the design group, and the requirements are never written. Under this type of organizational structure, it is difficult to hold anybody accountable for completing the software system.

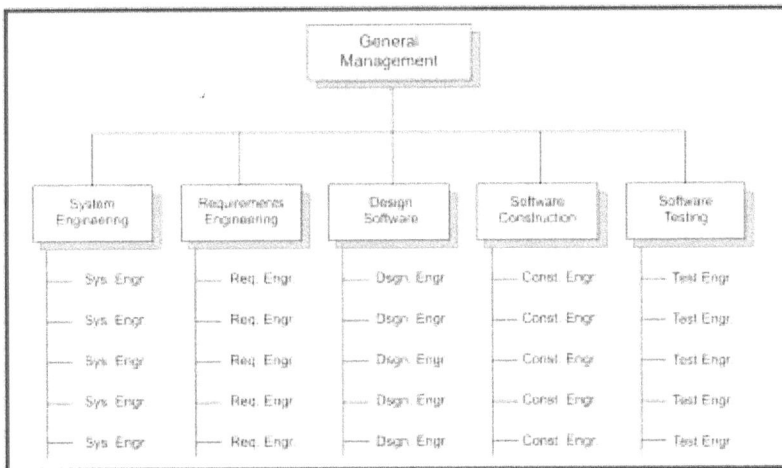

Figure 6: Functional organizational structure

b. ***Project organizational structure*** — A *project organizational structure* is typically used for very large, multipart, SW development projects. This type of software development project organization is built around a specific SW development project. Figure 7 portrays each project as a stand-alone organization with a project manager, and all assigned SW development skills. He/she is responsible for the completion and successful delivery of the software product.

The project manager is given the responsibility, authority, and resources for conducting the project. The manager must meet project goals adhering to allocated specified resources.

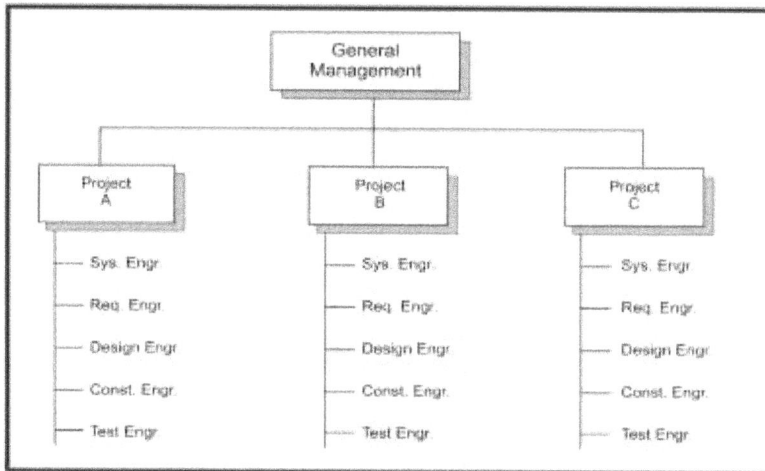

Figure 7: Project organizational structure

The project manager is responsible for completing the project on time and within scheduled costs. The manager is the liaison between the project and the customer. The project manager usually has the responsibility to hire, discharge, train, and promotes staff involved with project development and delivery.

c. ***Matrix organizational structure*** — Another type of software development project organization is the *matrix organization* (sometimes called a "matrix-project organization"), which is a compromise between the functional and project organizational structures. Matrix projects can be useful in any sized project. The project manager is given responsibility and authority for completing the project within the assigned resources. The functional managers provide the human resources needed for the project.

In a matrix organization, the project manager usually does not have the authority to hire, discharge, train, or promote personnel. As shown in Figure 8, each project is supervised by a project manager who daily oversees functional workers.

Typically, the software project manager is responsible for the day-to-day supervision of the software project members and the functional manager is responsible for the career, training, and well-being of project members. The functional manager has the responsibility and authority to hire, train, promote, and terminate software engineering personnel.

Since each individual worker is "supervised" by two separate managers—a project manager and a functional manager—the project organizational structure is sometimes called the "two boss system."

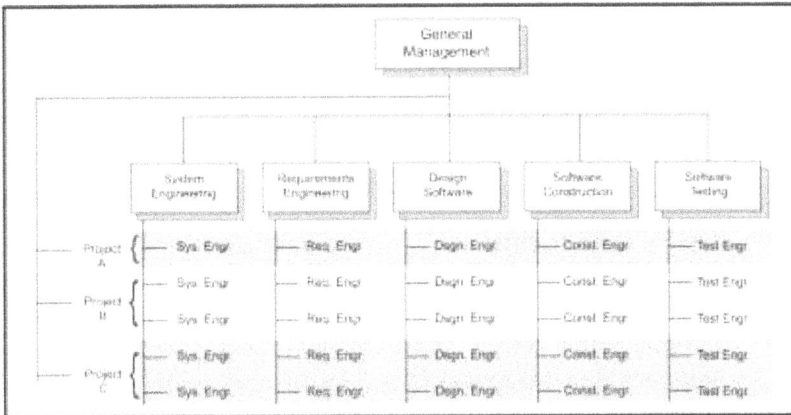

Figure 8: Matrix organizational structure

One of the major issues with the matrix project organizational structure is that it is too easy to move software engineers between projects—making it difficult for a project manager to know whom he or she can rely on when needed.

(3) **Software project teams.** A software team is normally a subset of a project structure. A software development project is typically organized around a number of SWE teams. These teams usually consist of five to fifteen members. Examples of structures for these teams include *egoless programming teams, chief programmer teams, and hierarchical teams* [Mantei 1981].

    a.   ***Egoless programming teams*** —The *egoless team* (a.k.a. a *democratic team*) structure was introduced by Dr. Gerald Weinberg in 1971. An egoless team typically consists of ten to fifteen members. Discussions and decisions are made by consensus. Reviews are called "egoless reviews" (now called "walkthroughs").

        Group leadership responsibility rotates; there is no permanent central authority. Egoless teams are not popular in the United States, which has a strong commitment to strong leadership, but they are used in other countries. Pure research teams are sometimes organized as *egoless teams*

    b.   ***Chief programmer teams*** — The *chief programmer team* (a.k.a. a *controlled centralized team*) was first used by IBM in the now-famous *New York Times Morgue Project* [Baker 1972]. The team consists of three or four permanently assigned team members—chief programmer, backup programmers, and program librarians—plus other auxiliary programmers and/or analysts who are added as needed.

The chief programmer manages all technical aspects and makes managerial and technical decisions related to the project. The librarian maintains all documents, code and data, and performs all administrative work. Chief programmer teams were popular throughout the 1970s and early 1980s. The title had a nice "ring" to it. The concept fell out of use due to a lack of skilled chief programmers.

c. ***Hierarchical teams*** — A *hierarchical structure* (a.k.a. a *controlled decentralized team*) is a structured organization in which the project leaders manage senior engineers and programmers; senior programmers and senior software engineers manage junior software engineers and programmers. The project team is called a "hierarchical team" because of its top-down flow of authority.

In today's environment, egoless and chief programming teams are seldom used. Therefore, the most commonly applied method is the hierarchical team (usually called a "project team"). However, the "walkthrough" (from the egoless programming team) and the "program librarian" (from the chief programmer team) are now included in many software development projects.

(4) **Unconventional teams.** Teams may have an unconventional structure, composed of members residing in different geographical regions or belonging to different product teams. *Virtual teams and cross-functional product teams* are two examples of unconventional teams.

a. ***Virtual teams*** — A virtual team does not really exist in the physical sense but exists in the logical sense. Team members have the same goals as a physical team, but they are geographically separated and tied together only by the use of communication lines such as e-mail. Although physically separated, team members may reside in Korea, the United States, Russia, Japan, and India, and are all connected, usually via the Internet.

Virtual teams try to perform like collocated teams. One of the *advantages* of a multinational virtual team is that team members reside in separate time zones. The ability of team members to work on a project 24 hours per day is an interesting phenomenon emerging from this structure. Team members in the United States arrive at work ready to proceed with a portion of the project that was forwarded during the night by team members in Russia.

A *disadvantage* is that the individual who has a particular piece of knowledge required to solve a specific problem may reside in a different time zone than where that knowledge is needed or that problem is being worked.

b. ***Cross-functional product teams*** — One reasonable approach to team organization requires a project to have one team, comprised of one member from each of the functional areas of the project. An example of this type of team might contain the following members: software engineer, hardware engineer, configuration manager, transportation manager, acquisition manager, and software quality-assurance manager. An Agile team is an important example of a cross-functional team.

3.6 Define responsibilities and authority.

Responsibility is the obligation to fulfill commitments. *Authority* is the right to make decisions and verify that they are put into practice. It is often stated that authority can be delegated but responsibility cannot be delegated. Koontz and O'Donnell [1972] support this view by defining responsibility as "the obligation owed by subordinates to their supervisors for exercising authority delegated to them in a way that accomplishes results expected." Authority for organizational activities or tasks should be assigned to the organizational position at the time it is created or modified. The project manager delegates the corresponding authority to the project organization within the project.

3.7 Establish position qualifications.

Position qualifications must be identified for each position in the project. Position qualifications are established by asking: What types of individuals do you need for your project? How much experience is necessary? How much education is required: B.S. in computer science, M.S. in artificial intelligence? How much training is required before or after the project is initiated? Does the applicant need to know Ada, C++, or some other programming language? The establishment of proper and accurate position qualifications will make it possible for the manager to staff the project correctly. Examples of typical position qualifications for SWE titles and positions are:

(1) ***Project managers*** — Background in successful systems implementation, advanced industrial knowledge, awareness of current computer technology, intimate understanding of user operations and problems, and proven management ability. Minimum requirements are four years of significant system development and project management experience.

(2) ***Software system engineers*** — Seven years of experience in aerospace applications designing real-time control systems for embedded computers. Experience with C++ preferred and a B.S. degree in computer science, engineering or related field.

(to be continued on Page 27)

3.8 Strengths and weaknesses of project organizational types.

Tables 6a-c display the advantages and disadvantages of the three organizational models—*functional, project,* and *matrix organizational models*

Table 6a: Strengths and weaknesses of functional organizations

Functional Organization	
Strengths	Weaknesses
Organization already exists (quick start-up and phase-down).	No central position of responsibility or authority for the project.
Easier recruiting, training, and retention of functional specialists.	Interface problems are difficult to solve.
Policies, procedures, standards, methods, tools, and techniques are already established.	Projects are difficult to monitor and control.

Table 6b: Strengths and weaknesses of project organizations

Project Organization	
Strengths	Weaknesses
There is a central position of responsibility and authority for the project.	An organization must be formed for each new project.
The organization provides central control over all system interfaces.	Career issues (recruiting, training, and retention of functional specialists) may be more difficult versus using the functional format.
Decisions can be made quickly.	Policies, procedures, standards, methods, tools, and techniques must be developed for each project.
Staff motivation is typically high.	Software developers are more prone to "bail out" during the concluding project phases of the project.

Table 6c: Strengths and weaknesses of matrix organizations

Matrix Project Organization	
Strengths	Weaknesses
Central position of responsibility and authority is stronger than in functional format.	Responsibility for and authority over individual project members is shared between two or more managers—unlike project or functional formats.
Interfaces between functions can be controlled more easily than in the functional format.	It is too easy to move personnel from one project to another—unlike project or functional formats.
Recruiting, training, and retention may be easier than in the project format.	More organizational coordination is required than in project or functional formats.
It is easier to start and end a project than in the project format.	There is greater competition for resources among projects than in the project or functional formats.
More flexible use of personnel is available than in the project or functional formats.	Individual software developers do not like the matrix organization because they can seldom advance to the level of project manager.
Policies, procedures, standards, methods, tools, and techniques are already established.	

3.7 Establish position qualifications (continued).

(4) *Scientific/Engineering programmers, programmer-analysts* — Three years of experience in programming aerospace applications, control systems, and/or graphics. In addition, one-year minimum experience using Ada, assembly, or C++ programming languages. Large-scale or mini/micro hardware exposure and system-software programming experience desired. Minimum requirements include undergraduate engineering or math degree.

(5) *Verification and validation engineer* — Minimum of three or more years of

experience in one or more aspects of V&V for real-time systems. Must be able to work independently of the development teams. M.S. degree in SWE preferred. Salary commensurate with experience.

(6) **Software quality assurance engineer** — Minimum of three years experience working in a software QA environment. Some CM experience desirable. B.S. or M.S. degree in computer science with specialty in software engineering.

4. Staffing a Software Engineering Project

The promotion of an outstanding engineer to project management, no matter how good an engineer he is, frequently results in gaining a poor manager and losing a good engineer. — RHT

Adding people to a late software project will just make it later — Brooks 1995.

4.1 Introduction and definitions.

Staffing a SWE project consists of all management activities that involve filling (and keeping filled) the positions that were established in the project organizational structure. This includes selecting candidates for each position and training or otherwise developing each candidate to accomplish his or her expected tasks. Staffing also involves terminating project personnel when necessary.

Staffing is not the same as organizing; staffing involves filling the roles created in the project organizational structure through selection, training, and development of personnel. The objective of staffing is to ensure that project roles are filled by qualified personnel (both technically and temperamentally) [Koontz & O'Donnell 1972].

4.2 Major issues in staffing actions.

The major issues in staffing for a SWE project are as follows:

(1) The productivity of programmers, analysts, and software engineers varies greatly among individuals.

(2) Project managers are frequently selected for their ability to program or perform engineering tasks rather than their ability to manage. Few engineers really make good managers.

(3) There is a high turnover of staff assigned to software projects, especially those organized under a matrix organization.

(4) Training plans for individual software developers are not produced or maintained. Thus, engineers may lack the specific skills needed for their current projects, as well as the general background skills required to enhance their career growth.

(5) Universities are not producing a sufficient number of computer science graduates who understand the SWE process.

In his book, *Software Engineering Economics*, Boehm [1981] reports a ratio of differences in productivity due to personnel and/or team capability as high as 4 to 1. The inability to predict accurately the productivity levels of individuals assigned to a project also undermines the software manager's ability to estimate the cost and schedule of software projects.

Today it is common practice to promote programmers and software engineers who have excelled at their technical activities to project managers. Unfortunately, success as a software developer (e.g., software engineer, programmer, and tester) does not always indicate a high potential to become a project manager. Compounding this problem is the lack of training in project management techniques and procedures made available to these budding project managers. Thus, management training needs to be improved.

Experience is a valuable commodity in a software development activity. Unfortunately, obtaining this experience is hampered by the constant turnover of project personnel. In the days of staff shortages, companies raided the personnel of their rivals, resulting in the movement of workers from one company to another and from one project to another. The use of the matrix organizational project format, discussed in the previous section, encourages the movement of software developers from one project to another as priorities change within a company. In addition, many software engineers recognize that the only way to receive a raise is to look for and accept a position with a different company (usually a 10% raise).

Every assigned job requires training in applied techniques and sciences and is a necessary element of staffing. Individual training plans ensure that training is available to meet the short- and long-term needs of the individual developer. Training plans are agreements between an individual developer and his/her manager detailing training (courses, seminars, study groups, or tuition support) that will be supported by the organization. In many, if not most organizations, these plans are not realized, leaving the pairing of individuals with appropriate courses to a rather ad hoc approach and further leaving the individual software engineer deficient in the skills needed to either complete the job or advance to another job.

Further, universities are not producing sufficient numbers of software engineers. Most of the computer science programs in U.S. universities are graduating theoretical computer scientists at best or merely programmers (coders) at worst. Most industry personnel and others involved with the hiring of new college graduates seek those with degrees from computer science programs that emphasize both theoretical education and practical experience in developing software systems—that is—*SWE skills.*

4.3 Staffing an SEPM project.

Table 7 provides an outline of the activities and tasks that must be accomplished by project managers when staffing their projects. *Note: SWEBOK does not include staffing as an SWE management technique.*

4.4 Fill organizational positions in a software project.

The project manager is responsible for filling the positions that were established during the organizational planning phase of the project. When staffing any software project, the following eight factors should be considered:

(1) *Education* — Does the candidate have the minimum level of education necessary to successfully complete assigned tasks? Does the candidate have the proper education required for future promotions in the company?

(2) *Experience* — Does the candidate have an acceptable level and length of experience? Is it the right type and variety of experience?

(3) *Training* — Is the candidate trained in the language, methodology, and equipment to be used and the software application area?

(4) *Motivation* — Is the candidate motivated to perform job assignments, work on the project, work for the company, and accept new assignments? *(College educated personnel are typically more adaptable to change than non-college graduates.)*

(5) *Commitment* — Will the candidate demonstrate loyalty to the project, to the company, and to the decisions made [Powell & Posner 1984]?

(6) *Self-motivation* — Is the candidate a self-starter, willing to carry a task through to the end without excessive direction.

(7) *Group affinity* — Does the candidate fit in well with the current staff? Are there potential conflicts that need to be resolved?

(8) *Intelligence* — Is the candidate able to learn, accept difficult assignments and adapt to changing environments?

Deficiencies in any of these factors can be offset by strengths in other areas. For example, deficiencies in education can be offset by greater experience.

4.5 Select a productive software staff.

The following metrics may indicate a productive software staff:

(1) *Amount of experience* — An experienced staff is more productive than an inexperienced staff [Boehm 1984]. Some of the most valuable experience comes from having worked on software projects similar to the project being staffed.

(2) *Diversity of experience* — Diversity of experience is a reasonable predictor of productivity [Kruesi 1982]. It is preferable that the individuals under consideration have performed well in several jobs over a period of time rather than in one job over a period.

(3) *Other qualities* — Other qualities indicative of a highly productive individual are effective communication skills (both oral and written), a col-

lege degree (usually but <u>not</u> necessarily in a technical field), being a self-starter, and experience in the application area of the project.

4.6 Assimilate newly assigned software personnel.

The manager is responsible not only for hiring project personnel but also for familiarizing them with any company or project policies, procedures, facilities, and plans necessary to ensure their effective integration into the project (i.e., "bringing them on-board"). In short, the project manager is responsible for introducing new employees to the company; and the company, to the employees.

Many large companies have formal orientation programs—often lasting several days or even weeks. When I worked for LMSC, Inc. the orientation programs include the features and history of the company, main sources of company revenue, general policies and procedures, organizational structure, company benefits, the availability of in-company service organizations, and the location of the nearest bathroom.

4.7 Educate or train personnel as necessary.

It is not always possible to recruit or transfer employees possessing the exact skills needed for a particular project. Therefore, the manager is responsible for training the assigned personnel to ensure that project requirements are met.

Education differs from training. Education involves teaching the basics, theory, and underlying concepts of a discipline with a view toward a long-term payoff. Training means teaching a skill or knowledge about how to use, operate, or create something. A training skill set typically has a short-term payoff.

For example, managers should be educated in management sciences and business techniques. They should be trained in management techniques and the duties of administration. Engineers, on the other hand, are educated in science and mathematics, but they must be trained in the application domain. All staff must be familiar with the procedures, tools, techniques, and equipment to be operated and used.

Software engineers need to be trained in the specific languages, operating systems, software development environments, tools, and so forth, that are used in the project.

Training methods include on-the-job training, formal company courses, courses taken through local universities and schools, self-study, computer-based training, and in-house lectures.

Each individual within an organization must have a training plan that specifies career, education, and training goals, and the steps each individual has taken and will take when achieving those goals. To be successful, top management must actively support training programs—project management is focused on the short term (the success of the project); higher management should focus on the long term (the success of the organization) and should understand the importance of training and education required for that objective.

Another technique is retraining for SWE (sometimes called "retreading") of longtime, valuable employees with obsolete skills or skills less in demand than previously needed [Ben-David et al. 1984, McGill 1984].

4.8 Provide for general development of the project staff.

In addition to *education* and *training,* the project manager must ensure that the project staff *grows* with both the project and the company. The manager must ensure that employees' professional knowledge will increase and that employees maintain positive attitudes toward the project, company, and customers.

One of the purposes of providing general development for the employee is to improve organizational effectiveness. For example, courses and degree programs at a local university in any relevant skill, funded by the company, will improve employee morale, aid in retaining employees, and broaden the skill base available to the company. Even indirect skills such as typing and communication should be enhanced.

4.9 Evaluate and appraise project personnel.

The project manager is also responsible for periodically *evaluating* and *appraising* personnel. Appraisals provide feedback to staff members concerning the positive and negative aspects of their performance. This feedback allows each staff member to strengthen effective qualities and improve those that are negative. Appraisals should be accomplished at regular intervals and should be concentrated on the individual's performance and not on personality traits—unless personality issues interfere with performance [Moneysmith 1984].

One well-known evaluation technique that is applicable to project management is "management by objectives" [Maslow 1954]. At the beginning of the appraisal period, the individual and the project manager establish a set of verifiable objectives that the individual believes can be obtained over the next reporting period. Together, these measurable objectives comprise a verifiable goal that forms the basis for the next appraisal.

This approach is superior to evaluation by personal traits and work characteristics, such as promptness, neatness, punctuality, golf scores, and so on. An example of management by objectives is the SMART (Specific, Measurable, Agreed, Realistic, Time-bound) approach to objectives originated by Hewlett-Packard.

4.10 Compensate the project personnel.

The manager—sometimes directly, sometimes indirectly—is responsible for determining the *salary scale* and *benefits* of project personnel. Benefits take on many forms. Most benefits are monetary or can be equated to money. These include stock options, a company car, first-class tickets for all company trips, or a year-end bonus.

Some benefits are nonmonetary but appeal to the self-esteem of the individual; examples are decorations and medals in the military, a promotion to a new position in the company (paid or not paid), a reserved parking place at the company plant, or an impressive office or title on the office door.

4.11 Terminate project assignments.

The project manager is not only responsible for hiring personnel but must also terminate assignments as necessary. "Terminate" includes reassignment of personnel at the end of a successful project (a pleasant termination) and dismissal of personnel due to project cancellation (an unpleasant termination).

Termination can also occur by firing when an employee is determined to be unsatisfactory [Davis 1997].

Termination is important. Poor performers not only fail to "pull their own weight;" they frequently present problems for the morale of others on the project. Management may be seen as ineffective if the poor performer is allowed to remain. Co-workers may feel resentful when another team member regularly shows up late to work or misses project deadlines.

An interesting termination occurs when a well-qualified engineer or manager is promoted to a new position with new responsibilities, and hopefully, a salary increase. While this is a "win-win" for the employee and the company, the promotion results in similar project termination issues as those resulting from losing the project.

5. Directing a Software Engineering Project

Trying to manage a team of software engineers is like trying to control a herd of cats. — *Sandy McGill, 1995*

5.1 Introduction and definitions.

Directing, like staffing, involves people. Directing is sometimes considered synonymous with leading (compare [Koontz & O'Donnell 1972] with [Koontz, O'Donnell & Weihrich 1984]). Directing involves providing leadership and day-to-day supervision of personnel, delegating authority, coordinating activities, facilitating communications, resolving conflicts, managing change, and documenting decisions.

Directing a SWE project consists of the management activities that involve *motivational* and *interpersonal aspects* by which project personnel come to understand and contribute to the achievement of project goals. Once subordinates are trained and oriented, the project manager has a continuing responsibility for clarifying assignments, guiding personnel toward improved performance, and motivating them to work with enthusiasm and confidence to accomplish project goals [Koontz & O'Donnell 1972].

5.2 Major issues in directing a software project.

The major issues in directing a SWE project are as follows:

(1) Companies and managers fail to communicate effectively between project and non-project entities.

(2) Companies and managers fail to understand that money is not a sufficient motivator for software developers.

(3) Companies and managers do not have the proper tools and techniques to motivate software engineers.

(4) Customers and managers do not recognize the potential software impact caused by a seemingly trivial SW change ("a simple matter of programming").

One of the major goals of SWE is to improve communication among the many organizations that are involved in developing a software system. Most SWE documents are written in a natural language (e.g., English), which is notoriously imprecise and ambiguous. Research in SWE is concerned with developing tools and techniques that will ease communication of requirements specifications, design documents, and other SWE documents.

Most software engineers are well paid, work in pleasant surroundings, and are satisfied with their position in life. According to Maslow's hierarchy of unfulfilled needs, the average software engineer is high on the ladder of satisfied needs [Maslow 1954].

Most software engineers reside at the "esteem and recognition" level, occasionally reaching to the "self-actualization" level. At this level, money alone is not a strong motivator. As a result, management is faced with the issue of how to motivate software engineers to produce more and better software (called "software psychology" in some circles).

The opportunity to use *modern tools* and *techniques* is a strong motivator for many software engineers. Technology transfer is defined as the time interval between the development of a new product, tool, or technique and its use in software development organizations. In their paper, Redwine and Riddle [1985], estimate that this time interval can span 5 to 18 years. The cause of this technology transfer gap can be examined from two viewpoints:

(1) The leadership team may be reluctant to introduce unfamiliar methods and tools because an introduction of these methods may increase risks to team projects. Higher management can mitigate this problem by recognizing that the project entails taking a risk on behalf of the company and providing additional resources or scheduling relief for the project.

(2) The use of unfamiliar methods and tools may make it more difficult for the leadership team to estimate project cost and schedule.

5.3 Directing an SEPM project.

Table 7 provides an outline of leadership activities and tasks that must be accomplished by project managers and team leaders.

Table 7: Directing activities for software projects

Activity	Definition or Explanation
Provide leadership	Create an environment in which project members can accomplish their assignments with enthusiasm and confidence.
Supervise personnel	Provide day-to-day instructions, guidance, and discipline to help project members fulfill their assigned duties.
Delegate authority	Allow project personnel to make decisions and expend resources within the limitations and constraints of their roles.
Motivate personnel	Provide a work environment in which project personnel can satisfy their psychological needs.
Build teams	Provide a work environment in which project personnel can work together toward common project goals. Set performance goals for teams as well as for individuals.
Coordinate activities	Combine project activities into effective and efficient arrangements.
Facilitate communication	Ensure a free flow of correct information among project members.
Resolve conflicts	Encourage constructive differences of opinion and help resolve the resulting conflicts.
Manage changes	Stimulate creativity and innovation to achieve project goals.

Software developers are discouraged by the seeming lack of understanding about the software development process on the part of managers and customers. SWE is one of the most difficult jobs in the world today. There are no small or easy software jobs. Software engineers who are under pressure to hurry up and finish that "simple change" become discouraged and may eventually "burn out" [Cherlin 1981].

5.4 Provide leadership to the project team.

The project manager provides leadership to the project team by ensuring that project activities are undertaken according to the project plan and schedule [SWEBOK

2004]. Throughout the process, resources are utilized (for example, personnel effort and funding) and deliverables are produced (for example, architectural design documents and test cases).

Leadership entails implementing project plans, interpreting plans and requirements, and ensuring that everyone on the project team is working toward common goals. Leadership is a function that results from the power of the leader and his or her ability to guide and influence individuals. The project manager's power can be derived from his or her leadership position; this is called "positional power." The project manager's power can also be derived from his or her own "charm." This type of power is called "personal power."

A good leader is able to align the personal goals of subordinates with the organizational goals of the project. Problems can arise when the project manager, who has only positional power, comes into conflict with a subordinate who has personal power over the project members [Boyatzis 1971].

Either the project manager or his or her team will prepare and execute agreements with suppliers and monitor supplier performance [SWEBOK 2004].

5.5 Supervise project personnel.

The project manager is responsible for overseeing work completed by project members and providing day-to-day supervision of the personnel assigned to the project. It is the project manager's responsibility to provide guidance and, when necessary, to discipline project members.

Supervisory responsibilities can involve such mundane tasks as clocking in the employees at the beginning of the workday, approving vacation time, reprimanding an individual for a missed appointment, or approving a deviation from company policy. At other times, the project manager can make a crucial decision with regard to a software design approach, propose a well-reasoned argument to upper level management that results in procurement of better tools and work space, or be a sympathetic listener to a project member's personal problems.

5.6 Delegate authority to the appropriate project members.

The SWE project manager is also responsible for delegating authority to the project staff. Tasks are assigned to subgroups, teams, and individuals, and authority is delegated to teams, enabling them to accomplish tasks in an efficient and effective manner. Typically, a good project manager will always delegate authority to the staff member responsible for the lowest possible manager in the project. Decisions should be made at the lowest possible level (and are usually best made at this level) [Raudsepp 1981].

The proper delegation of the right kind of authority can free managers from time-consuming routine supervision and decisions, thus enabling them to concentrate on the important aspects of the project.

The project manager should ensure that individual project members understand the authority delegated for each responsibility. Project members should clearly understand the scope, limitations, and purpose of the delegation.

5.7 Motivate project personnel.

The project manager is responsible for motivating and inspiring personnel to execute their best work. Several motivational techniques from mainstream management are applicable to SWE projects, such as management by objective, Maslow's hierarchy of needs [Maslow 1954] and Herzberg's hygiene factors [Herzberg, Mausner, & Snyderman 1959].

The motivational models and techniques listed in Table 8 have been developed over the past 50 years and should be familiar to most project managers.

The project manager should always acknowledge the special needs of the highly qualified, technically trained engineers and scientists who staff the project. *Money will attract good software engineers to a company; money will not retain them* [http://tech beacon.com/4-ways-recruit-retain-software-engineers].

Total quality management (TQM), McGregor's theory X and Y, and job factors for computer personnel are three examples of several leading motivation techniques:

(1) ***Total quality management (TQM)*** — TQM can be thought of as a motivation technique applicable to an organization or enterprise. TQM is a strategy for continuously improving performance at each level and area of responsibility. TQM focuses on processes that create products. By delegating the authority to make decisions and improve quality to the lowest level, the TQM process provides a degree of motivation to the personnel implementing it. TQM is based on [Deming 1986].

(2) ***McGregor's Theory X and Y*** — McGregor presented two theories concerning human nature, called Theory X and Theory Y [McGregor 1960]. It should be noted that contrary to popular belief, McGregor did not favor one view over the other. He did not say that Theory Y presented a better management technique than Theory X. McGregor only proposed the two theories.

   a. ***Theory X assumptions*** — Human beings possess an inherent dislike of work and will avoid it if possible. Because of this dislike of work, most people must be coerced, controlled, directed, and threatened with punishment to encourage them to put forth adequate effort toward the achievement of organizational objectives. Human beings prefer to be directed, wish to avoid responsibility, have relatively little ambition, and desire and strive for security above all other factors.

   b. ***Theory Y assumptions*** — The expenditure of physical and mental effort towards work is as natural as is play or rest. External control and the threat of punishment are not the only means for achieving organizational objectives. People will exercise self-direction and self-control in the attainment of objectives to which they are committed. Commitment to objectives is a function of the rewards associated with achievement. Motivated human beings will not only accept responsibility but also seek it.

Table 8: Motivation models and techniques

Motivation Models	Definition or Explanation
Frederick Taylor	Workers will respond to a wage incentive.
Elton Mayo	Interpersonal (group) values override individual values.
Kurt Lewin	Interpersonal (group) values override individual values. Group forces can overcome the interests of an individual.
Douglas McGregor	Managers must understand people in order to be able to motivate them.
Abraham Maslow	Human needs can be categorized in a hierarchy. Satisfied needs are not motivators.
Frederick Herzberg	A decrease in environmental factors is dissatisfying; an increase in environmental factors is not satisfying. A decrease in job content factors is not dissatisfying; an increase in job content factors is satisfying.
Chris Argyris	The greater the disparity between company needs and individual needs; the greater the dissatisfaction of the employee.
Rensis Likert	Participative management is essential to personal motivation.
Arch Patton	Executives are motivated by the challenge of work, status, the urge to achieve leadership, the lash of competition, fear, and money.
Total quality management (TQM)	A strategy for continually improving performance at each level and area of responsibility.

(3) *Job factors for computer personnel* — Table 9 provides a list of factors, in order of declining importance, that motivate computer personnel to accept a job (left-hand column) and a list of factors that cause a job to be unattractive or discouraging to the job-holder; that is, reasons for either not accepting or for quitting a job (right-hand column) [Fairley 2009].

One of the cardinal rules of project management states that, on any team with five or more team members, being *the team leader (manager) is a full-time job; the team leader should not be expected to carry a share of the "technical" load.*

Having to accomplish technical work in addition to management tasks means that the management function must be limited to day-to-day supervision and "firefighting" (solving immediate problems), with no time devoted to important activities such as longer-term planning.

Table 9: Job factors for software personnel

Attractive Factors	Discouraging Factors
Salary	Company mismanagement
Chance to advance	Poor work environment
Work environment	Little feeling of accomplishment
Location	Poor recognition
Benefits	Inadequate salary
Facilities/equipment	Poor chance to advance
Job satisfaction	Poor facilities/equipment
Company management	Poor benefits
Job responsibility	Poor career path definition

5.5.8 Build software development teams. A project team is typically a sub-organization of a SWE project. This section of the chapter extends the discussion of team types covered in Paragraph 5.3.5(3) and sources of software engineers discussed in Paragraph 5.4.4 [http://www.pluralsight.com].

(1) Project teams should number between 5 and 15 engineers and a team leader. Small teams are generally more efficient than large teams; they can be easier to lead and can contribute more significantly to the project goals.

(2) Each team should be led by a capable software engineer who has demonstrated the leadership and capacity to lead a group of software engineers.

(3) Team leaders are not "project managers," in that they do not have the greater responsibilities to hire, discharge, train, or promote personnel.

(4) Using a team that is "jelled" [DeMarco & Lister 2013] and enjoys working together is much more important for most projects than "herding" a bunch of egotistical geniuses who do not get along. Lasting software is built simply and you want to hire the smartest people you can, but temper that urge by hiring people who will fit with your team.

5.9 Coordinate activities.

Coordinate activities to combine the different elements of a complex activity or organization into a relationship that will ensure efficiency and harmony. Any group working together toward a common goal needs an individual (normally a team leader, project manager, or someone with the authority of a manager) to ensure that all pieces and parts of project elements fit together and interface with all outside elements and functions.

5.10 Facilitate communication.

Along with coordination, the project manager is responsible for facilitating communication both within the project and between the project and other organizations—to facilitate means to expedite, ease, and assist in the progress of communication. "Communication" is used here to mean the exchange of information among entities that are working toward common goals.

For example, when practical, the project manager should disseminate the staffing plans and project schedule throughout the organization. Nothing can destroy the morale of an organization faster than false and misleading rumors. A good project manager will ensure that the project staff is kept well informed in order to quickly dispel rumors.

5.11 Resolve conflicts.

It is the project manager's responsibility to resolve conflicts among project staff members and between the staff and outside agencies in both technical and non-technical matters. The project manager is not expected to be an expert in all aspects of the project, but should possess sound judgment and problem solving skills.

The project manager should reduce the likelihood of future conflict by removing potential sources of disagreement whenever possible (e.g., team members with equal positions should have equal benefits, accesses to the manager, parking spaces, and so forth).

The project manager should also notate possible conflict between the employee's work activities and personal life. When such conflict reaches epic proportions, it is called "burnout" [Cherlin 1981].

5.12 Manage change that affects the software project.

The project manager is responsible for encouraging independent thought and innovation in achieving project goals. A good manager must always accommodate change when change is cost-effective and beneficial to the project [Kirchof & Adams 1986].

It is important for the project manager to control change and not discourage changes that are cost effective. Clearly, requirements, design, and the application area for which the software system is built will change. There will be social changes. What is acceptable to build at one time will not necessarily be acceptable later. People change; newly graduated engineers will enter the workforce with innovative ideas as they have been taught modern methods of developing software systems. The bottom line is not to eliminate change but instead to control it. Yourdon [1987] presents a simple, systematic plan for the transfer of a new software technology (a change) into a software development organization:

(1) Explain the risks and benefits of the new method, tool, or technique.

(2) Provide training for the project team.

(3) Prototype the technique before it is used.

(4) Provide technical support throughout the project.

(5) Listen to the users' concerns and problems.

(6) Avoid concentrating on the technology at the expense of the project.

Staff turnover is another type of change, one in which negative consequences generally receive the most attention. Below is a listing of the major negative *and* positive consequences of turnover applicable to SWE projects:

(7) *Negative consequences of staff turnover*:

    a. *Major project disruption* — The loss of key personnel during software development can seriously delay and even permanently impair important projects.

    b. *Loss of strategic opportunities* — Turnover of critical personnel can sometimes cause organizations to postpone or cancel projects that would have otherwise enhanced their position relative to competitors and/or increased profits.

    c. *Recruitment and selection costs* — The monetary costs associated with recruitment and selection involves such costs as advertising, employment agency fees, travel costs associated with campus visits or recruitment at computer-related associations, entertainment costs for prospective recruits, administrative costs, and even bonus payments payable to employees for recommending new prospects.

    d. *Training and development costs* — Even if the individual possesses the necessary technical skills, he or she will experience a learning curve associated with the process of becoming familiar with the organization. Turnover thus incurs these costs as well.

e. ***Decline in morale*** — Personnel leaving an organization will frequently cause morale problems, particularly if the individual is perceived as leaving due to poor working conditions.

f. ***Breakup of a cohesive team*** — The departure of even a single individual may have serious detrimental effects on a cohesive work group (i.e., team), at least in the short term..

(8) ***Positive consequences of staff turnover:***

a. ***Increased performance*** — A positive consequence is that relatively poor performers will leave, to be replaced by more skilled performers.

b. ***Salary and benefit cost savings*** — Higher levels of turnover normally produce a workforce with less longevity. A relatively junior work force is much less costly to the organization in the form of regular pay, overtime pay, FICA costs, pensions, and length of vacations.

c. ***Innovation and adaptability*** — New employees can bring technological expertise and experience gained from other companies and organizations.

d. ***Increased internal mobility*** — Particularly when senior employees depart, turnover enables promotional possibilities for high-performing lower-level staff.

e. ***Increased morale*** — When poor performers or individuals who tend to create destructive conflict leave, the result can include increased morale.

(9) ***Quality-replacement matrix*** — Negative and positive consequences of turnover are normally viewed in terms of their cumulative effects on an organization (e.g., high turnover is costly or low turnover does not provide new and innovative ideas, fresh views, and stimulation). Yet, when one considers actually managing turnover, the perspective must shift to turnover as an individual phenomenon. Decisions concerning whether or not an individual remains with an organization are products of the forces to remain versus the forces to leave.

The goal of the organization is to retain those individuals who are high-quality employees in terms of job skills, particularly those who are difficult to replace. Similarly, those employees of inferior quality who are easily replaced should be terminated a.s.a.p.

6. Controlling a Software Engineering Project

Planning and controlling are like the blades on a pair of scissors. The scissors cannot work unless both blades are there.

—Unknown Manager

6.1 Introduction and definitions.

Controlling is the collection of management activities used to ensure that the project proceeds according to plan. Controls provide plans and approaches for eliminating the differences between plans and standards, and actuals and results.

Control is a feedback system that provides information detailing how well the project is progressing. Whereas directing is very people-focused, controlling is concerned with numbers: actuals versus planned. Performance and results are measured against plans, deviations are noted, and corrective actions are taken to ensure conformance of plans and actuals.

Control asks the following questions:

(1) Is the project on schedule?

(2) Who is responsible for assessing progress? Who will take action to analyze reported problems?

The control process also requires organizational structure, communication, and co-ordination.

Methods and tools must be objectively controlled. Information must be quanti-fied. The methods and tools must point out deviations from plans without re-gard to the particular people or positions involved. Control methods must be tailored to individual environments and managers. The methods must be flexi-ble and adaptable to the changing environment. Control also must be economi-cal; the cost should not exceed project benefits.

Control issues must lead to corrective action—for example, either to bring the actu-al status back to the current plan, to change the plan, or to terminate the project. Control is the project manager's primary responsibility during that portion of the project life cycle in which the system or software is actually being built. It is essen-tial that deviations from plans, budgets, and schedules be detected and corrected in a timely fashion; otherwise, a high risk of complete project failure exists.

6.2 Major issues in controlling.

Major issues in controlling a SWE project are as follows:

(1) There is a tendency to use budget expenditures to measure "progress" without considering the amount of work accomplished.

(2) Progress in a software development project is difficult to visualize and measure.

(3) Quality is frequently not required, monitored, or controlled.

(4) Often standards for software development and project management are not written or, if written, are not enforced.

(5) The body of knowledge called "software measurements" or "metrics"

(used to measure the productivity, quality, and progress of a software product) has not been fully developed.

A reliance on budget expenditures for the management of progress is a major issue related to controlling a software project. For example, when a project manager is asked for the status of a software project, he or she will typically analyze the resources expended. If three-quarters of project funds have been expended, the project manager will report that the project is three-quarters completed. The obvious problem here is that the relationship between resources consumed and work actually accomplished is merely a rough measure at best and completely incorrect at worst.

Very few methods for measuring scheduled progress are accurate and easy to use. The "earned-value method" and "binary tracking system" are two of the best metrics used to monitor software systems; however, they are time consuming and costly to implement [Howes 1984].

Software engineering standards that can be used for measuring progress and products are often not written or, if written, are not enforced. Process standards are sometimes considered detrimental to software projects because they "stifle creativity." Therefore, it is possible for project managers to ignore company standards in favor of local, ad hoc, and frequently inadequate project control systems.

Another issue related to control of a software engineering project arises with regard to users and buyers of software systems that do not specify quality in their software system or in their request for proposals (RFPs). Project managers do not feel obligated to build a quality product if the customer has not requested it, enforced it, or specifically funded it.

Issues related to control of a software project arise primarily because the body of knowledge known as "software metrics" is not fully developed. However, project managers are not proficient with "a priori" design of software systems that, when implemented, will have the desired quality attributes. This deficiency has resulted in emphases being placed on the *processes* of SWE in the belief that sound development processes will result in high-quality *products*.

The bad news is that budget expenditures are still used as the primary metric of progress in many SWE projects.

6.3 Controlling an SEPM project.

Table 10 provides an outline of the project management activities that must be accomplished by project managers to fully control each phase of their projects. At critical points in the project, overall progress toward achievement of the stated objectives and satisfaction of stakeholder requirements are evaluated. Similarly, assessments pertaining to the effectiveness of the overall process to date, the personnel involved and the tools and methods employed are undertaken at particular milestones (SWEBOK 2004).

Table 10: Controlling activities for software projects

Activity	Definition or Explanation
Develop standards of performance	Set goals that will be achieved when tasks are accomplished.
Establish monitoring and reporting systems	Determine necessary project data, who will receive data, when data will be received, and what will be done to control the project.
Measure and analyze results	Compare achievements with standards, goals, and plans.
Initiate corrective actions	Bring requirements, plans, and actual project status into conformance with the project plans.

6.4 Develop standards of performance.

A *standard* is a documented set of criteria used to specify and determine the adequacy of an action or object. A *SWE standard* is a set of procedures that define the process for developing a software product and/or that specify the quality of a software product.

The *project manager* is responsible for developing and/or specifying applicable standards of performance for the project. The project manager either develops standards and procedures for the project, adopts and uses standards developed by the parent organization, or uses standards developed by the customer or a professional society. (*See for example, IEEE Software Engineering Standards.*)

Process standards and *product standards* are both important in developing high-quality software. SWE is primarily concerned with the process of developing software rather than with the measurement of the product. Software quality metrics (measuring such quality attributes as software reliability, maintainability, portability, and other "-ilities") is not a well-developed science. Tools and techniques that effectively measure the quality of a software product are generally not available. Most SWE standards involve the process of *developing* software. In addition to providing a gauge by which to measure the SWE process, SWE standards also offer a substantial advantage for SWE organizations to measure software quality. Standards include the following:

(1) Improved communication among team members.

(2) Easier transfer of staff among projects.

(3) Expedited sharing of project experiences and history.

(4) Reduced need to retrain engineers, designers and programmers between projects.

(5) Ability to apply the best experiences from successful projects.

(6) Simplified software implementation and software maintenance.

(7) Improved control of projects since the standard for each particular process can be controlled.

(8) Ability to apply software quality assurance to every phase of the project.

6.5 Establish monitoring and reporting systems.

Software methods, procedures, tools, and techniques must also be specified, monitored, and reported. Tools that aid in the control of a SWE project are PERT, CPM, workload charts, and Gantt charts. The following thirteen monitoring and reporting systems are used by the project manager to track a SWE project: ADR, milestone review, binary tracking, work product specification, V&V, testing, software configuration management, and software quality assurance.

(1) **Architectural design review (ADR)** — An *ADR* is sometimes incorrectly called a "preliminary design review" or a PDR. It is held at the completion of the architectural design phase.

(2) ***Milestone review*** — *Milestone reviews* are usually chaired by either the customer or higher-level management. The project manager presents the status and progress, the work completed to date, the budget expended, the current schedule, and any management and technical problems that may have surfaced since the last review. The review is a success when either the customer or top level management gives permission for the project manager to proceed to the next phase.

(3) ***Earned value tracking*** — The paper authored by Howes [1984] presents the *earned-value* method of tracking a SWE project. The purpose of earned value tracking is to:

    a. Determine critical functions or resources to be tracked—normally resources or costs.

    b. Allocate the resource budget to individual elements of the work breakdown structure.

    c. Allow the allocated budget for an element of the WBS to be "earned back" when that element is completed. Partial completion is not permitted; the task is either completed or it is not. This is called "binary tracking."

    d. Determine the status of total costs to date:

        i. If (earned value < actual expenditures) then the project is over budget.

        ii. If (earned value < budget-to-date) then the project is behind schedule.

(4) **Work product specification** — A *work product specification* (a.k.a. work package specification) is a hieratical description of the tasks (work) to be accomplished when completing a function, activity or task. A work package specifies the objectives of the work, staffing, the expected duration, resources, results, and any other special considerations included in the work project. Work packages are normally small tasks that can be assigned to two or three individuals to be completed in 2-3 weeks. A series of work packages comprises a software project.

(5) **"98% completion syndrome"** — The "*98% completion syndrome*" is a cynical view of the method of determining project status by examining the project as a whole and trying to determine what percentage is complete. For example, given a 6-week project, an immature project team will report that the project is 98% complete in the final week of the project with only "one or two bugs to find and fix." The project manager continues to report 98% completion for the next six weeks or more, long after its estimated due date, until the project is finally completed. The issue is that the project manager was unaware of the actual project status.

(6) **Binary tracking** — *Binary tracking* is a solution to the "98% completion syndrome. " In binary tracking, the concept is that a work package is either complete or not complete (i.e., assigned a numeric "1" or "0"). Binary tracking of work packages is a reasonably accurate means of tracking the completion of a software project. For example, if a 6-week project has 880 work packages and 440 are completed at the end of the fifth week, the project is 50% complete, not 98% complete. Binary tracking is also a major support tool belonging to the earned value concept. (*See [Howes 1984] for a succinct description of the earned value concept.*)

(7) **Unit development folder** — The *unit development folder* (UDF) (sometimes called the "software development folder") is a specific form of development notebook that has proven to be useful and effective in collecting and organizing software products as they are produced. The purpose of the UDF is to provide an orderly approach to the development of a program unit and to provide management visibility and control over the development process [Ingrassia 1987].

(8) **Walkthrough and inspection** — *Walkthroughs* and *inspections* are reviews of a software product (design specifications, code, test procedures, etc.) conducted by the peers of the group being reviewed [Ackerman 1996]. Walkthroughs are a critique of a software product by the producer's peers for the sole purpose of finding errors. The inspection system is another peer review developed in 1976 by Michael Fagan [1976] of IBM. Inspections are typically more structured than walkthroughs.

(9) **Independent auditing** — The *independent audit* is an independent review of a software project used to determine compliance with software requirements, specifications, baselines, standards, policies, and software quality assurance plans.

An independent audit is an audit performed by an outside organization not associated with the project [Bernstein 1981]. On the positive side, an independent team provides an unbiased opinion of the project's development. In the case of a need for expert knowledge, the independent team can supplement an existing team skill set. A negative aspect of the independent audit is that the audit team needs to be well versed with every aspect of project development. To be effective, an audit team must be continuously involved with the project.

(10) **Verification and validation** — *Verification and validation* is one of the most capable methods of determining whether the product is correctly developed. Verification ensures that each phase of the life cycle correctly interprets the specification from the previous phase. Validation ensures that each completed software product satisfies its requirements [Fujii & Wallace 1996].

(11) **Testing** — *Testing* is the controlled exercise of the program code in order to expose errors. The four primary types of testing are unit, integration, system, and acceptance. *Unit testing* is the testing of one unit of code (usually a module) by the programmer who programmed the unit. *Integration testing* is the testing of each unit or element in combination with other elements to show the existence of errors between the units or elements. *System testing* is the controlled exercise of the completed system. *Acceptance testing* is the "demonstration" to the customer or buyer that the to-be-delivered software meets its constructional requirements. In each case, it is important to develop test plans, test procedures, test cases, and test results [IEEE Std. 829-2008].

(12) **Software configuration management** — *Software configuration management* (SCM) is a method for controlling software documentation and reporting software status. SCM is the discipline of identifying the configuration of a system at discrete points in time. This is done for the purpose of systematically controlling changes to this configuration and maintaining integrity and traceability throughout the system life cycle [Bersoff 1984].

(13) **Software quality assurance** — *Software quality assurance* (SQA) is "a planned and systematic pattern of all actions necessary to provide adequate confidence that the item or product conforms to established technical requirements" [IEEE Std. 729-1983]. SQA includes the development process and management methods (requirements and design), stand-

ards, configuration management methods, review procedures, documentation standards, verification and validation, and testing specifications and procedures. SQA is one of the major control techniques available to the project manager.

6.6 Measure and analyze results.

The project manager is responsible for *measuring* the results of the project, both during and at the end of the project. For instance, actual phase deliverables should be measured against planned phase deliverables. The measured results can be management (process) results and/or technical (product) results. The status of the project schedule is an example of a process result. An example of a product result is the degree to which the design and methods are used to measure results. . The following four paragraphs describe the measurement of product result design and methods:

(1) **Quality and quantity metrics** — A *metric* is a measure of the degree to which a process or product possesses a given attribute. Process and product metrics and other definitions of metrics are adapted from [IEEE Std. 610.12-R2002].

(2) **Software quality metrics** — A SW *quality metric* is a measure of the degree to which software possesses a given attribute that affects its quality. Examples are reliability, maintainability, and portability [IEEE Std. 1061-1998].

(3) **Software quantity metrics** — A SW *quality metric* is a measure of a physical attribute of software. Examples are lines of code, function points, and pages of documentation.

(4) **Management metrics** — A *management metric* is an indicator used to measure management activities such as budget spent, value earned, cost overrun, and schedule delays [AFSC Pamphlet 800-43 1986].

Measurement is an integral part of the Software Engineering Institute's Process Improvement Project. The first set of guidelines published by this project [Humphrey & Sweet 1987] included 10 metrics at maturity level 2:

a. Planned vs. actual staffing profiles

b. Software size vs. time

c. Statistics on software code and test errors

d. Actual vs. planned units designed

e. Actual vs. planned units completed during unit testing

f. Actual vs. planned units integrated

g. Target computer memory utilization

h. Target computer throughput utilization

i. Target computer I/O channel utilization

j. Software build/release content

With the 1993 publication of the Software Engineering Institute's Capability Maturity Model for Software [Paulk et al. 1993], measurements became one of the "common features" that are part of every key process area.

6.7 Initiate corrective actions for the project.

If standards and requirements are not being met, the project manager must *initiate corrective action*. For instance, the project manager can change the plan or standard, use overtime or other procedures to get back on track, or change the requirements (e.g., deliver fewer products (software) than initially specified).

The project manager might change the plans or standards if it is apparent that the original plans or standards cannot be met. This might involve requiring a larger budget, increased staff or personnel, and additional testing on the development computer. It also might require reducing the standards (and indirectly the quality) by reducing the number of walkthroughs, limiting review of all software modules, or only reviewing the critical software modules.

It is possible to return to the schedule by increasing resources. (However, this is not always the case. Brooks's Law states that "adding people to a late project can make it later [Brooks 1995].) The project manager must revise the project plan to account for the additional resource requirements. It is also possible to maintain the original cost by lengthening the schedule and/or reducing the functionality of the software system. In this situation, the plan must be revised to show the change in requirements or schedule.

7. Summary

SWE procedures and techniques alone do not guarantee development of a successful project. A poor manager struggles with every problem, real or imaginary—and no rule, policy, standard, or technique will help to resolve problems that arise during the development process under the watch of an inexperienced manager. But a good project manager can sometimes overcome or work around deficiencies in line organizations, staffing, budgets, standards, or other elements contained in a SWE project. The methods and techniques discussed in this article, in the hands of a competent project manager, can improve the probability of delivering a successful project.

In this chapter and in many other documents, the terms "project management" and "SWE project management" are used interchangeably. This is because the management of a SWE project and other types of projects require many of the same tools, techniques, approaches, and methods used in mainstream management. The functions and general activities of management are the same at all levels; only the detailed activities and tasks differ.

> **The probable cause of most software project failures is attributed to ineffective or nonexistent software engineering project management.**

Appendix A

Appendix A provides additional descriptions of issues present in federal, state, and private sector software engineering projects

A.1 Department of Defense software problems.

The importance of software project management is best illustrated in the following paragraphs extracted from the cited Department of Defense (DOD) reports:

(1) A report from the STARS (Software Technology for Adaptable, Reliable Systems) initiative states, "The manager plays a major role in software and systems development and support. The difference between success or failure—between a project being on schedule and on budget or late and over budget—is often a function of the manager's effectiveness" [DoD Software Initiative 1983].

(2) A Report to the Defense Science Board Task Force on Military Software states that today's "major problems with software development are not technical problems, but management problems" [Brooks 1987].

(3) A General Accounting Office (GAO) report that investigated the cost and schedule overrun of the C-17 states, "Software development has clearly been a major problem during the first 6 years of the program. In fact, the C-17 is a good example of how not to manage software development." [GAO/IMTEC-92-48 C-17 Aircraft Software 1992].

Capable project management is the single most important activity of any software engineering project.

A.2 State of California potential software problems.

When it comes to finances, California's vast state government often operates in the form of dozens of smaller government entities. Many use separate computer systems developed years ago—decades even—to record and process everything from payroll and procurement to caseload management and document tracking. The systems are not uniform, which causes the most basic task such as employee payroll and vendor billing to be a complex and labor-intensive system.

To improve its operation, the State has attempted to automate many of its activities. As a result, the State has launched dozens of software development projects—many of them mammoth undertakings. Moreover, many (I hesitate to say "most") of them were project failures. The State employs very few dedicated software engineering project managers, and as a result, it relies on contractors to not only build the SW systems but to also manage the systems.

To the State's credit, the State of California recognized the value of verification and validation (V&V) and effectively applied it to several design projects. The State's latest solution is to develop a system that will replace the aging and nonintegrated financial systems with a single comprehensive financial application supporting the fiscal and policy decision processes.

The proposed solution, called the *Enterprise Resource Planning* application, is designed to meet the state's budget, accounting and some procurement needs. The solution will also address various fiscal information reporting requirements of the Legislature. The project was initiated in July 2005 and is scheduled for completion in 2017, 11.9 years after it started, at a total cost of $617 million, of which $122 million has already been spent [Sacramento Bee 2013b].

A.3 Other software crisis issues.

(1) Several years ago, a paper written by Wayt Gibbs appeared in *Scientific American*, titled "Software Chronic Crisis." This paper revolves around the issue of commercial systems that had experienced a "software crisis" because the custom-built software either could not be delivered on time and within budget, or did not work. Examples of so-called software failures cited in this particular paper by Gibbs were the baggage control system at the Denver International Airport and a driver's license and vehicle registration system for the State of California Department of Motor Vehicles [Gibbs 1994].

(2) Recently, the U.S. government health insurance program—the *Affordable Care Act* (*ACA*) (Obamacare)—sign-up system was initiated. Many state exchanges crashed and the probable causes were that the systems were not built—or required—to handle the rush of individuals who wanted or needed health insurance. The Republican Party interpreted this as a "sign" that the American people did not want "Obamacare." However, we know better—that this failure revealed the symptoms of a poorly managed SW system project.

Reasons for the SW failure given by the U.S. Government's General Accounting Office (GAO) were as follows [GAO 2014]:

a. The responsible agency, the Center for Medicare and Medicaid Services (CMS), adopted insufficient project planning and oversight practices.

b. The project period was compressed (i.e., there was not enough time to properly test the system before launch).

   c.   The CMS used a cost-reimbursement type of contract.

   d.   The CMS did not use available software quality assurance (SQA) plans to perform government oversight.

   e.   Meeting project deadlines was the driving factor (versus quality).

   f.   The adopted incremental information technology development approach (Agile) was new to the CMS.

   g.   There were frequent requirements changes.

   h.   Contract task orders were issued before the task requirements were available.

   i.   A risk-analysis strategy was not prepared.

   j.   CMS launched the ACA system without verifying that the system could meet performance system requirements.

(3) In 2005, an article in *IEEE Spectrum* by Robert N. Charette, listed a number of reasons why software engineering systems fail so often [Charette 2005]. Most failures are caused by management mistakes:

   a.   Unrealistic or unarticulated project goals (a planning issue).

   b.   Inaccurate estimates of needed resources (a planning issue).

   c.   Badly defined system requirements (a planning or system engineering issue).

   d.   Poor reporting of the project's status (a controlling issue).

   e.   Poor communication among customers, developers, and users (a directing issue).

   f.   Use of immature technology (a planning issue).

   g.   Inability to handle the project's complexity (an organizing or staffing issue).

   h.   Inadequate development practices (a planning issue).

   i.   Poor project management (a self-defined issue).

   j.   Stakeholder politics (a planning or risk analysis performance issue).

   k.   Commercial pressures (a planning or risk analysis performance issue).

Appendix B

The purpose of adding this "ancient" document to this chapter is to remind readers that software engineering has not changed in the last 40 plus years.

The software development policies listed below were established by TRW, a well-known and respected system engineering company. In the early 1970s TRW discovered that it did not have an "in-house" policy for developing *software systems*. This was particularly troublesome because TRW was famous for taking other software developers "to task" for poor software development procedures.

To make amends, senior TRW managers established a corporate policy to initiate and maintain a set of software development processes to be used for in-house software development. (This process preceded the popular use of the term "software engineering.") The set of processes, titled the *Software Development Policy*, was developed by the TRW SEID. (SEID was the acronym for the TRW software development organization.)

The reason for exhibiting the table of contents for the TRW software engineering processes listed below is to enforce the point that software engineering has not changed materially since at least 1976. The two major differences in software engineering in 1976 versus software engineering today (2017) is that "primary" design is now "architectural" design and "critical" design is now "detailed" design.

SEID SOFTWARE DEVELOPMENT POLCIES

1 January 1977

The following is a list of the SEID Software Development Policies:

3.4.1   Software Requirements Specification

3.4.2   Software Requirements Review and Acceptance

3.4.3   Software Specification (Preliminary)

3.4.4   Software Preliminary Design Review

3.4.5   Software Design Specification (Critical)

3.4.6   Software Critical Design Review

3.4.7   Top Down Software Design

3.4.8   Unit Development Folders

3.4.9   Software Design Walk-Throughs

3.4.10 Programming Standards

3.4.11 Software Unit Test Plan

3.4.12 Software System Integration and Test

3.4.13 Software Acceptance Test Plan and Procedures

3.4.14 Software User's Manual

3.4.15 Software End-Product Acceptance Plan

3.4.16 Software Documentation

3.4.17 Software Configuration Management

3.4.18 Software Quality Assurance

REFERENCES

Additional information about the SEPM KA can be found in the following references:

- **[Ackerman 1996]** Frank A. Ackerman, "Software Inspections and the Cost-Effective Production of Reliable Software," in *Software Engineering,* M. Dorfman and R.H. Thayer, eds. IEEE Computer Society Press, Los Alamitos, CA, 1996.

- **[AFSC Pamphlet 800-43 1986]** "Air Force Systems Command Software Management Indicators: Management Insight," *AFSC Pamphlet 800-43,* HQ AFSC Andrews AFB, Washington, D.C., 31 January 1986.

- **[Baker 1972]** F. Terry Baker, "Chief Programmer Team Management of Production Programming," *IBM Systems Journal,* Vol. 11, pp. 56-73, 1972.

- **[Bartol & Martin 1983]** K.M. Bartol and D.C. Martin, "Managing the Consequences of DP Turnover: A Human Resources Planning Perspective," in *Proceedings of the 20th ACM Computer Resources Planning Perspective,"* ACM, New York, 1983, pp. 79-86.

- **[Ben-David et al. 1984]** A. Ben-David, M. Ben-Porath, J. Loeb and M. Rich, "An Industrial Software Engineering Retraining Course: Development Considerations and Lessons Learned," *IEEE Transactions on Software Engineering,* Vol. SE-10, no. 1, Nov. 1984, pp. 748-755.

- **[Bernstein 1981]** L. Bernstein, "Software Project Management Audits," *Journal of Systems and Software,* Elsevier, North Holland, 1982, pp. 281-287.

- **[Bersoff 1984]** E.H. Bersoff, "Elements of Software Configuration Management," *IEEE Transactions on Software Engineering,* Vol. SE-10, no. 1, January 1984, pp. 79-87.

- **[Boehm 1981]** B.W. Boehm, *Software Engineering Economics,* Prentice-Hall, Englewood Cliffs, NJ, 1981.

- **[Boehm 1984]** B.W. Boehm, "Software Engineering Economics," *IEEE Transactions on Software Engineering,* Vol., SE-10, no. 1, January 1984.

- **[Boyatzis 1971]** R.E. Boyatzis, "Leadership: The Effective Use of Power," *Management of Personnel Quarterly,* Bureau of Industrial Relations, 1971, pp. 1-8.

- **[Brooks 1987]** "Report on the Defense Science Board Task Force on Military Software," Office of the Undersecretary of Defense for Acquisition, Department of Defense, Washington, D.C., September 1987.

- **[Brooks 1995]** F. Brooks, *The Mythical Man Month: Essays on Software Engineering,* 2nd ed., Addison-Wesley, Upper-Saddle River, 1995.

- **[Buckley 1987]** F.J. Buckley, "Establishing Software Engineering Standards in an Industrial Organization," in *Tutorial: Software Engineering Project Management,* R.H. Thayer, IEEE Computer Society Press, Los Alamitos, CA, 1988.

- **[Charette 2005]** Robert N. Charette, "Sep 2," *IEEE Spectrum,* 2005.

- **[Cherlin 1981]** M. Cherlin, "Burnout: Victims and Avoidances," *Datamation,* July 1981, pp. 92-99.

- **[Cleland & King 1972]** David I. Cleland and William R. King, *Management: A Systems Approach,* see "Table 5: Major Management Functions as Seen by Various Authors," McGraw-Hill Book Company, N.Y., 1972.

- **[Davis 1997]** A. Davis, "Trial by Firing: Saga of a Rookie Manager," in *Software Engineering Project Management,* IEEE Computer Society Press, Los Alamitos, CA, 1997.

- **[DeMarco & Lister 2013]** Tom DeMarco and Tim Lister. *Peopleware: Productive Projects and Teams,* 3rd ed., Addison-Wesley, Reading, MA 2013.

- **[Deming 1986]** W. Edwards Deming, *Out of the Crisis,* MIT Center for Advanced Engineering Study, Cambridge, MA, 1986.

- **[Dijkstra 1972]** Edsger Dijkstra, "The Humble Programmer" (EWD340), *Communications of the ACM,* Vol. 15, no. 10 (Oct. 1972).

- **[DoD Software Initiative 1983]** *Strategy for a DoD Software Initiative,* Department of Defense Report, October 1, 1982. (An edited public version was published in *Computer,* November 1983.)

- **[Donnelly, Gibson, & Ivancevich 1975]** J.H. Donnelly, Jr., J.L. Gibson and J.M. Ivancevich, *Fundamentals of Management: Functions, Behavior, Models,* Rev. ed., Business Publications, Dallas, TX, 1975.

- **[Fagan 1976]** M.E. Fagan, "Design and Code Inspections to Reduce Errors in Program Development," *IBM Systems Journal,* Vol. 15, no. 3, 1976, pp. 182-211.)

- **[Fagan 1986]** M.E. Fagan, "Advances in Software Inspections," *IEEE Transactions on Software Engineering,* Vol. SE-12, no. 7, July 1986, pp. 744-751, CA, 1988.

- **[Fairley 2009]** Richard E. (Dick) Fairley, *Managing and Leading Software Projects,* IEEE Computer Society, Wiley, Hoboken, NJ, 2009.

- **[Fayol 1949]** H. Fayol, *General and Industrial Administration,* Sir Isaac Pitman & Sons, Ltd., London, 1949

- **[Fitz-enz 1978]** J. Fitz-enz, "Who Is the DP Professional?" *Datamation*, September 1978, pp. 125-128.

- **[Fujii & Wallace 1996]** Roger Fujii and Dolores R. Wallace, "Software Verification and Validation," in *Software Engineering*, M. Dorfman and R.H. Thayer, IEEE Computer Society Press, Los Alamitos, CA, 1996.

- **[GAO 2014]** GAO Report: "Ineffective Planning and Oversight Practices Underscore the Need for Improved Contract Management," Government Accountability Office, GAO-694, July 2014.

- **[GAO/IMTEC-92-48 C-17 Aircraft Software 1992]** GAO Report: "Embedded Computer Systems: Significant Software Problems on C-17 Must Be Addressed," *General Accounting Office GAO/IMTEC-92-48*, Gaithersburg, MD 20877, May 1992.

- **[Gibbs 1994]** W.W. Gibbs, "Software's Chronic Crisis," *Scientific American*, Sep. 1994, pp. 86-95.

- **[Herzberg, Mausner, & Snyderman 1959]** F. Herzberg, B. Mausner and B.B. Snyderman, *The Motivation to Work*, John Wiley & Sons, New York, 1959.

- **[Howes 1984]** N.R. Howes, "Managing Software Development Projects for Maximum Productivity," *IEEE Transactions on Software Engineering*, Vol. SE-10, no. 1, January 1984, pp. 27-35.

- **[Humphrey & Sweet 1987]** W.S. Humphrey and W.L. Sweet, "A Method for Assessing the Software Development Capability of Contractors," CMU/SEI-87-TR-23, Sep. 1987.

- **[IEEE Software Engineering Standards]** A list of these standards can be found at [http://www.ieeec.com/publications_standards/publications/sub scriptions/prod/standards_overview.html].

- **[IEEE Std. 1012-2004]** *IEEE Standard for Software Verification and Validation,* IEEE, New York, 2004.

- **[IEEE Std. 1061-1998]** *IEEE Standard for a Software Quality Metric Methodology,* IEEE, New York, 1998.

- **[IEEE Std. 610.12-R2002]** *IEEE Glossary of Software Engineering Terminology,* IEEE, Piscataway, NJ, 2002.

- **[IEEE Std. 729-1983]** *IEEE Standard Glossary of Software Engineering Terminology*, IEEE, New York, 1983.

- **[IEEE Std. 730-2002]** *IEEE Standard for Software Quality Assurance Plans,* IEEE, New York, 2002.

- **[IEEE Std. 829-2008]** *IEEE Standard for Software Test Documentation*, IEEE, New York, 2008.

- **[Ingrassia 1987]** F.S. Ingrassia, "The Unit Development Folder (UDF): A Ten-Year Perspective," in *Tutorial: Software Engineering Project Management*, edited by R.H. Thayer, IEEE Computer Society Press, Los Alamitos, CA, 1988.

- **[Kirchof & Adams 1986]** N.S. Kirchof and J.R. Adams, "Conflict Management for Project Managers: An Overview," extracted from *Conflict Management for Project Managers*, Project Management Institute, February 1986, pp. 1-13.

- **[Koontz & O'Donnell 1972]** H. Koontz and C. O'Donnell, *Principles of Management: An Analysis of Managerial Functions*, 5th ed., McGraw-Hill Book Company, New York, 1972.

- **[Koontz, O'Donnell, & Weihrich 1980]** H. Koontz, C. O'Donnell and H. Weihrich, *Management*, 7th ed., McGraw-Hill Book Co., New York, 1980.

- **[Koontz, O'Donnell, & Weihrich 1984]** H. Koontz, C. O'Donnell and H. Weihrich, *Management*, 8th ed., McGraw-Hill Book Co., New York, 1984.

- **[Kruesi 1982]** Betsy Kruesi, "Seminar on Software Psychology," California State University, Sacramento, Fall 1982.

- **[Los Angeles Times 2013]** Chris Megerian, "California Fires Contractor on Tech Project," *Los Angeles Times,* Feb. 8, 2013.

- **[MacKenzie 1969]** R.A. MacKenzie, "The Management Process in 3-D," *Harvard Business Review*, Vol. 47, no. 6, Nov.–Dec. 1969, pp. 80-87.

- **[Mantei 1981]** M. Mantei, "The Effect of Programming Team Structures on Programming Tasks," *Communications of the ACM*, Vol. 24, no. 3, March 1981, pp. 106-113.

- **[Maslow 1954]** A.H. Maslow, *Motivation and Personality*, Harper & Brothers, New York, 1954.

- **[McGill 1984]** J.P. McGill, "The Software Engineering Shortage: A Third Choice," *IEEE Transactions on Software Engineering*, Vol. SE-10, no. 1, Jan 1984, pp. 42-48.

- **[McGregor 1960]** D. McGregor, *The Human Side of Enterprise*, McGraw-Hill, New York, 1960.

- **[Mil.-Std. 498-1995]** Military Standard: *Software Development and Documentation*, Department of Defense, Dec. 5, 1995.

- **[Moneysmith 1984]** M. Moneysmith, "I'm OK – and You're Not," *Savvy*, April 1984, pp. 37-38.

- **[Neumann 1993]** P.G. Neumann, "System Development Woes," *Communications of the ACM*, Vol. 36, no. 10, Oct. 1993, pp. 146.

- **[Paulk et al. 1993]** Mark C. Paulk, Bill Curtis, Mary Beth Chrissis and Charles V. Weber, "Key Practices of the Capability Maturity Model, Version 1.1," CMU/SEI-93-TR-25, February 1993.

- **[Paulk et al. 1996]** Mark C. Paulk, Bill Curtis, Mary Beth Chrissis and Charles V. Weber, "The Capability Maturity Model for Software," in *Software Engineering*, M. Dorfman and R.H. Thayer, IEEE Computer Society Press, Los Alamitos, CA, 1996.

- **[Powell & Posner 1984]** G.N. Powell and B.Z. Posner, "Excitement and Commitment: Keys to Project Success," *Project Management Journal*, 15(4), 1984, pp. 30–46.

- **[Raudsepp 1981]** Eugene Raudsepp, "Delegating Authority." IEEE Press, New York, 1981.

- **[Redwine & Riddle 1985]** Samuel T. Redwine and William E. Riddle, "Software Technology Maturation," *Proceedings of the Eighth International Conference on Software Engineering*, IEEE Computer Society Press, Los Alamitos, CA, August 1985, pp. 185-200.

- **[Reifer 2006]** Donald J. Reifer, *Software Management*, 7th ed., IEEE Computer Society Press, Los Alamitos, CA, 2006

- **[Rue & Byars 1983]** Laslie W. Rue and Lloyd L. Byars, *Management: Theory and Application*, Richard D. Irwin, Inc., Homewood, IL, 1983.

- **[Sacramento Bee 1997a]** A State Automated Child Support System was developed at a cost of $100M. After eight years, the project was canceled.

- **[Sacramento Bee 1997b]** A Correctional Management Information System was initiated to manage the state's prisoners and parolees, at a cost of $44M. The V&V contactor said it could not be built from the design document. The State sued the software development company for $25M and won, *Sacramento Bee*, 1997.

- **[Sacramento Bee 2013a]** Unemployment Insurance Payment Processing System that would allow claimants to submit required certifications for jobless benefits online or by telephone.

- **[Sacramento Bee 2013b]** *Enterprise Resource Planning*, designed to meet the state's budget, accounting and some procurement needs, *Sacramento Bee*, 2013.

- **[Sacramento Bee 2015a]** "A State Automated Child Support System was developed at a cost of $100M. After eight years, the project was canceled, *Sacramento Bee*, 2015.

- **[Sacramento Bee 2015b]** From 1994 to 2013, California State government spent $985 million on seven computer projects that either were terminated or suspended, *Sacramento Bee*, 2015.

- **[Standish Group Report, 2006]** *Standish Group Report*, 2006.

- **[Standish Group Report, 2008]** *Standish Group Report*, Web page accessed 11-23-2008.

- **[SWEBOK 2004]** *Guide to the Software Engineering Body of Knowledge*, IEEE, New York, 2004.

- **[Thayer 1988]** R.H. Thayer, ed., *Tutorial: Software Engineering Project Management*, IEEE Computer Society Press, Los Alamitos, CA, 1988.

- **[Thayer 2000]** R.H. Thayer, ed., *Software Engineering Project Management,* 2nd ed., IEEE Computer Society Press, Los Alamitos, CA, 2000

- **[Thayer & Pyster 1984]** R.H. Thayer and A.B. Pyster, "Guest Editorial: Software Engineering Project Management," *IEEE Transactions on Software Engineering*, Vol. SE-10, no. 1, January 1984.

- **[TRW 1977]** *SEID Software Development Policies*, TRW, 1977.

- **[Weihrich 1993]** Heinz Weihrich and Harold Koontz, Chapter 1: *Management: A Global Perspective*, 10 ed., McGraw-Hill, Inc, NY, 1993.

- **[Weinberg 1971]** G. Weinberg, *The Psychology of Computer Programming,* Van Nostrand Reinhold, New York, 1971.

- **[Youker 1977]** R. Youker, "Organizational Alternatives for Project Management," *Project Management Quarterly*, Vol. VIII, no. 1, March 1977, pp. 18-24.

- **[Yourdon 1987]** E. Yourdon, "A Game Plan for Technology Transfer," in *Tutorial: Software Engineering Project Management*, R.H. Thayer, IEEE Computer Society Press, Los Alamitos, CA, 1987.

Chapter 2
Principles of Software Engineering
Project Management[2]

Donald J. Reifer
Reifer Consultants, LLC

ABSTRACT

This paper communicates 14 principles of software engineering project management that are based upon the experience of seasoned software managers. To make these principles useful, each is related to the primary functions that software managers perform. These principles are based on the fundamental premise that good engineering and classical project management methods, tools, and techniques can be applied in a cost-effective manner to cope with the challenges associated with delivering high-quality software products on schedule and within budget.

1. Introduction

This article introduces you to 14 principles of software management that revolve around the *five primary functions* that software project managers perform to get their job done effectively: *planning, organizing, staffing, directing,* and *controlling*. Software managers must develop skills, knowledge, and abilities in each of these functions to successfully deliver an acceptable product on schedule and within budget. They start by planning their projects thoroughly and creating the road map used to create baselines and expectations. Then they create organizations, staff them with the right mix of talent and capability, develop teams and teamwork, and motivate and direct their human resources toward achieving project-related goals. They have to integrate the work of many participants to pull the pieces of this puzzle together in such a manner that aggressive budgets and schedules can be achieved.

Software managers put controls in place and use them to track project status and determine whether teams are making suitable progress. They manage risk and deal with the day-to-day issues that can, if left unchecked, impede progress. They replan, refocus, and reenergize their teams as they take detours to overcome the obstacles that get in the way of achieving goals. Software project managers are focused on delivering acceptable products on

---

2. Extracted from Donald J. Reifer, *Software Management*, 7th edition, John Wiley & Sons, Hoboken, NJ, 2006. Used with permission of the author.

schedule and within budget. They intentionally avoid doing things that would distract them from accomplishing this goal.

2. Software Management Tutorial (7th edition) Organization

This article discusses the organization of the seventh edition of the *Software Management* tutorial and provides you with an overview of its contents. The tutorial is organized into ten chapters and three appendices. The first three chapters provide the necessary background for the tutorial, acquainting the reader with the topics of software life cycle models and process improvement. This chapter introduces you to the discipline of software project management and its tools.

The five sections that follow discuss six primary functions of software engineering project management: planning, organizing, staffing, directing, controlling, and instituting technology change.

3. Planning

Planning is defined as deciding in advance what has to be done, when and how to do it, and who should do it. It encompasses many related disciplines, such as estimating, budgeting, and scheduling. Software managers get involved in many types of planning exercises. For example, they plan projects, capital acquisitions and/or training, and skill development. Plans form the basis against which schedule and budgetary performance are assessed and project control is implemented. Plans create the foundations that project managers use to gain visibility into and control over progress. Based upon the experience of seasoned managers and the software project management body of knowledge [PIMBOK 2000], the following three principles establish the basis for project planning.

- *Principle 1: Planning Takes Precedence.* Planning logically takes precedence over all other management functions. While often difficult and time-consuming to perform, plans form the basis for all future work. Managers are encouraged to devote the time needed to figure out what needs to be done, when to do it, who should do it, and how to address the contingencies. Budgets are financial plans, whereas schedules create a viable project timeline.

- *Principle 2: Effective Plans Tap the Infrastructure.* Plans are most effective when they are consistent with policies and take full advantage of the organization's existing management infrastructure. This is especially true for those organizations that have initiated a software process improvement program aimed at institutionalizing a "preferred process" and "best practices" across groups.

- ***Principle 3: Plans Should Be Living Documents.*** Plans should be maintained as living documents or they will quickly become outdated and lose their value as control tools. Plans need to be periodically updated to add detail and reflect the current p r o j e c t   s t a t u s .

Because of their short timetable, most project plans tend to be tactical (near-term) instead of strategic (long-term). Allowing capital and research budgets to be impacted is a mistake because implementation of project plans may become dependent on other agencies for their realization. The most basic elements of a project plan are its budgets (financial plans) and schedules (delivery timetables). The parts of these documents with the highest leverage are the risk management and contingency plans.

It is not uncommon for a manager to spend as much as fifty percent of his/her time on planning during the early project phases. The better the plans, the more visibility and control you have over task progress. In addition, the higher in management you move, the more strategic your plans become. For example, product line or line-of-business managers generate product plans whose horizons may be 10 to 20 years long. Independent of the level of planning, each plan you develop represents a guide to some future course of action.

To establish a budget, you will have to prepare a resource estimate (time, staff, budget, etc.) based upon your understanding of the work that must be done and the resources available to complete the job. The ability to estimate resources accurately is a skill every software manager must possess. Poor resource estimates can lead to problems that no amount of dedication, perseverance, and hard work can correct.

4. Organizing

Managers create organizations to achieve their goals and complete the work they are responsible for as efficiently as possible. Such organizations provide a structure that lets managers complete work by assigning responsibilities, delegating authority, and holding people accountable for results. Most managers work within an existing organizational structure. Their function is to build teams, staff them, provide them with leadership, direct them, integrate their results, and manage communications up, down, and across the organization. Based upon extensive software engineering project management experience, these tasks lead to the following two organizational principles:

- ***Principle 4: Assign Your Software Manager Early.*** Recruit your software manager early during the project and empower this professional to perform all of the tasks for which he or she will be held responsible. Ensure that this person occupies a high enough position in the hierarchy to successfully compete for needed resources (staff, budget, etc.), talent, and management support.

- *Principle 5: Give Authority Commensurate with Responsibility.* Your software manager's responsibility should be commensurate with his/her authority. Because software managers are not always masters of their own destiny, they should not be held accountable for results when others' actions impede their performance. Such problems often occur when dealing with requirements. Software managers are not in charge of requirements, but rely on them to form the foundation of their architecture and design.

Many of the organizational factors that impact the performance of software managers fall outside their sphere of control. For example, marketing is often responsible for developing requirements (and their frequent change) and for customer liaison. Influence is the key for gaining control over this untenable situation. The software manager must be able to communicate effectively with others within, across, and up and down the existing organizational framework. Working groups and use of integrated product teams are mechanisms that can be used specifically for this purpose.

Communications must flow up and down and across the organization to keep people informed. Staff members must keep abreast of current events or they will lose focus and their performance will be negatively impacted. Newsletters, colloquiums, brown-bag lunches, weekly team meetings, and monthly "all hands" meetings, are proven mechanisms for improving communications. They should be exploited, along with software peer reviews and inspections.

5. Staffing

Staffing refers to recruiting, appraising, growing, and keeping the people you need to properly complete and deliver the project. Organizations are only as good as the people who populate them. Software managers must be able to recognize talent, breed competence, and weed out deadwood. They must also be able to attract the right people to fill key slots within their organizations. Based upon a great deal of software engineering project management experience, the following two principles establish the basis for staffing:

- *Principle 6: Care about Your People.* Software managers must be able to show their staff that they truly care about them, their careers, and their goals. Because actions speak louder than words, managers must demonstrate their devotion by fighting for promotions, salary increases and better working conditions for their people. They must also be able to coach poor achievers and be able to improve their job-related performance.

- ***Principle 7: Provide Dual-Career Ladders.*** Promotion should be possible through either a technical or managerial career ladder. Technical people who do not want to move into management slots should be given the opportunity to follow other career paths. Chief software engineer positions that are equivalent to middle- and upper-level management slots should be a visible part of the organizational chart.

In many organizations, dual-career paths act as powerful incentives for technical people who may or may not want to move into management. Knowing what is required to progress along dual lines provides software specialists with growth and career opportunities. Knowledge acts as a motivator and stimulates high levels of achievement. It also makes career counseling easier when staff are not satisfied.

As in many other scientific disciplines, good technical performers are often promoted prematurely to management positions. This is frequently a mistake because the skills required for management differ from those required for engineering. Good software managers have to be developed. Training must be provided for those who have demonstrated management potential. Mentoring and other forms of coaching need to be available to help develop individual skills and abilities. In addition, new supervisors should be taught the fundamentals of management.

6. Directing

Managers accomplish assignments through the actions of others. They communicate their goals and lead and motivate their subordinates to achieve these goals, typically under deadline pressures. Direction tends to be difficult because software developers and engineers are highly creative and individualistic. Leadership and direction are needed to eliminate mistrust and provide focus for work activities. Based upon lessons learned managing software projects, the following three direction principles create a foundation for our discussions about direction:

- ***Principle 8: Provide Your People with An Opportunity to Excel.*** Interesting work and the opportunity to excel will motivate your staff to accomplish project goals. Software managers need to understand how to channel behavior so that it is directed towards achieving work-related goals.

- ***Principle 9: Lead by Satisfying Goals.*** People will follow those individuals who lead by example and represent a means to satisfy their own goals. Success will come to those managers who can make satisfying personal and work-related goals.

- **Principle 10: Keep Key People Focused.** Avoid giving your best people too much to do. Too many diversions cause a loss of efficiency that talent alone cannot correct. Learn to say "no" to distractions. Keep your people focused on the current task.

We would like to populate our organizations with talented, self-motivated professionals. Under such a system, tasks would be completed with little or no management interference. Unfortunately, such situations do not exist in most firms. Managers must build synergistic teams and motivate their players to perform to their fullest capability. Managers must be able to communicate, to lead, and motivate players so they can survive the trials of combat. In addition, they must be able to focus the team, as necessary, to meet deadlines.

7. Controlling

Planning and controlling tend to be closely related activities. Managers control by tracking progress against plans and acting on observed deviations. They track actuals against targets and forecast trends. Controls should be diagnostic, therapeutic, accurate, timely, understandable, and most importantly, economical. They should call attention to significant deviations from the norm and suggest ways of fixing the problems. They should be forward-looking and emphasize what is needed in the future to make corrections relative to plans.

Controls should be imposed throughout the software development process. To be in control, managers must manage risks and use metrics to manage. Based upon extensive project management experience, this leads to the following three control principles:

- **Principle 11: Focus on The Significant Deviations.** Controls should be implemented to promptly alert managers to significant deviations from plans. The philosophy of "if it isn't broke, don't fix it" should be remembered. In other words, don't interfere if things are going well and the prognosis looks good for the future.

- **Principle 12: You Cannot Control What You Cannot Measure.** Effective control requires that we measure performance against standards. Normally, these standards are budgets and schedules established as targets within the project plan. However, other standards may exist, especially when metrics and other indicators are used to track status and measure progress. Independent of the system used, you cannot determine where you are going if it is not known where you are or have been.

- **Principle 13: Make Risk Abatement Your Goal.** Risk management and abatement must be an integral part of any control system or it will cease to work. Identifying obstacles and figuring out ways to avoid them in advance is an essential part of the control process.

Controls close the loop in the feedback system. They provide managers with the visibility and insight needed to make better and timely decisions. As noted in Reifer's book, *Software Management* [2006], software managers rely on configuration management, metrics and measurement, quality assurance, software inspections, risk management, and verification and validation techniques to provide them with visibility into the project's status and control over its progress.

8. Instituting Technology Change

The software industry is in a constant state of change. Managers need to be aware of advances that are being made in order to harness them for their benefit. Although using new technology may sometimes be risky, software managers must be able to figure out when and how to apply technology for their organization's benefit. Otherwise, the ability to complete the project may be hindered. Some of the emerging management concepts are again discussed [Reifer 2006]. Experience with such new technology gives rise to the following final principle of software project management:

- ***Principle 14: Match Technology Risk with Expected Benefits.*** Technology should be used only when the risk associated with its use is acceptable. For projects on a tight schedule, the introduction of something new may be unacceptable. Yet, the same technology may be defensible when used on another project where adequate resources are available to insert it operationally.

I firmly believe that technology transfer is the primary means we have to alleviate most of the software problems the industry is experiencing. We need to figure out how to tap the benefits of new technologies, like those explained in my book, without paying too high a price. We need to work smarter and harder, or we may not be able to handle the workload in the future.

9. Summary and Conclusions

I am indebted to many good managers with whom I have had the pleasure of working over the past thirty years. They have taught me a great deal. Their conduct has influenced my conduct. Their wisdom has made me wiser. Their experience has become a part of mine. They have provided me with role models, mentored and coached me, spurred me on when I needed motivation, and influenced my management style.

My goal with this tutorial is to help you improve your management capabilities by communicating the lessons I have learned primarily from others in the form of the 14 principles of software management that I used to organize and shape my volume [Reifer 2006].

I would like to thank the IEEE Computer Society for motivating me to keep my tutorial current. The field of software management has made great strides since the last edition of this tutorial. I hope the next few years will see us make even more progress.

10. Final Thoughts

It is interesting to reflect back 27 years when my tutorial was first published. At that time, I was teaching project management and could not find a suitable text. I wrote the first edition to fill that gap. Professors of those universities and colleges that adopted this book as their text told me that they also could not find a suitable textbook. Now, when I ask my professor friends why they are still using my text, they give a different reason. They say that there are too many texts available. They use this tutorial because it distills the body of knowledge in software project management to the point where it can be taught. It is nice to see such progress.

REFERENCES

- **[PIMBOK 2000]** *A Guide to the Project Management Body of Knowledge,* Project Management Institute, Newtown Square, PA, 2000.

- **[Reifer 2006]** Donald J. Reifer, *Software Management,* 7th edition, IEEE Computer Society Press, Los Alamitos, CA, 2006.

Chapter 3
Management: Science, Theory and Practice[3]

Heinz Weihrich

University of San Francisco

One of the most important human activities is managing. Ever since people began forming groups to accomplish aims they could not achieve as individuals, managing has been essential to ensure the coordination of individual efforts. As society has come to rely increasingly on group effort and as many organized groups have grown larger, the task of managers has been rising in importance. The purpose of this book is to promote excellence of all persons belonging to organizations, but especially managers, aspiring managers, and other professionals.

1. Definition of Management: Its Nature and Purpose

Management is the process of designing and maintaining an environment in which individuals, working together in groups, accomplish efficiently selected aims. This basic definition is expanded by the following points:

(1) As managers, people carry out the managerial functions of planning, organizing, staffing, leading, and controlling.

(2) Management applies to any kind of organization.

(3) Management applies to managers at all organizational levels.

(4) The aim of all managers is the same: to create a surplus.

(5) Managing is concerned with productivity; that implies effectiveness and efficiency.

1.1 The universality of management.

All managers carry out the functions of planning, organizing, staffing, leading, and controlling, although the time spent in each function will differ and the skills required by managers at different organizational levels vary. Still, all managers are engaged in getting things done through people. Although the managerial concepts, principles, and theories have general validity, their application is an art and depends on the situation. Thus, managing is an art using the underlying sciences. Managerial activities are common to all managers, but the practices and methods must be adapted to the particular tasks, enterprises, and situations.

---

3. This paper has been modified from Chapter 1 of *Management: A Global Perspective*, 10th ed., edited by Heinz Weihrich and Harold Koontz, New York: McGraw-Hill, 1993. Used with permission of the author.

This concept is sometimes called the *universality of management* in which managers perform the same functions regardless of their place in the organizational structure or the type of enterprise that they are managing.

Management applies to any kind of organization and applies to managers at all organizational levels. Management also applies to small and large organizations, to profit and not-for-profit enterprises, and to manufacturing as well as service industries. The term "enterprise" refers to businesses, government agencies, hospitals, universities, and other organizations, because almost everything said in this book refers to business as well as nonbusiness organizations. Effective managing is the concern of the corporation president, the hospital administrator, the government first-line supervisor, the Boy Scout leader, the bishop in the church, the baseball manager, and the university president.

The universality of management concept makes no distinction between managers, executives, administrators, and supervisors. Managers are charged with the responsibility of taking actions that will make it possible for individuals to make their best contributions to group objectives. To be certain, a given situation may differ considerably among various levels in an organization or various types of enterprises. Similarly, the scope of authority held may vary and the types of problems may be considerably different.

Furthermore, the person in a managerial role may be directing people in the sales, engineering, or finance department. Nevertheless, the fact remains that, as managers, all obtain results by establishing an environment for effective group endeavors.

The universality of management *concept* allows us to apply all general management functions "across-the-board" to *all* types of management. The universality of these concepts provides a management framework for adapting traditional management functions to project management. From these general management functions, this chapter derives the detailed activities and tasks that should be undertaken by a manager assigned to a software engineering project.

1.2 The functions of management.

Many scholars and managers have found that the analysis of management is facilitated by a useful and clear organization of knowledge. As a first order of knowledge classification, we have used the five functions of managers: *planning, organizing, staffing, leading,* and *controlling*. Thus, management concepts, principles, theory, and techniques are organized around these functions.

This framework has been used and tested for many years. Although there are different ways of organizing managerial knowledge, most textbook authors today have adopted either this or a similar framework even after experimenting at times with alternative ways of structuring knowledge.

Although the emphasis in this article focuses on managers' tasks in designing an internal environment for performance, it must never be overlooked that managers must operate in the external environment of an enterprise as well as in the internal environment of an organization's various departments. Clearly, managers cannot perform their tasks well unless they understand and are responsive to, the many elements of the external environment—economic, technological, social, political, and ethical factors that affect their areas of operations.

Managerial functions provide a framework for organizing management knowledge. All new management ideas, research findings, or techniques can be placed into the functions of planning, organizing, staffing, leading, and controlling.

1.2.1 Planning. *Planning* involves selecting missions and objectives and the actions to achieve them; it requires decision making—that is, choosing future courses of action from among alternatives. There are various types of plans, ranging from overall purposes and objectives to the most detailed actions to be taken, such as to order a special stainless steel bolt for an instrument or to hire and train workers for an assembly line. No real plan exists until a decision—a commitment of human or material resources or reputation—has been made. Before a decision is made, all we have is a planning study, an analysis, or a proposal, but not a real plan.

1.2.2 Organizing. *Organizing* involves people working together in groups to achieve a desired goal. Some groups must have roles to play much like the parts actors fill in a drama, whether these roles are ones they develop themselves, are accidental or haphazard, or are defined and structured by someone who wants to make sure that people contribute in a specific way to a group effort.

The concept of a "role" implies that what people do has a definite purpose or objective; they know how their job objective fits into group effort and they have the necessary authority, tools, and information to accomplish the task.

Organizing, then, is that part of managing that involves establishing an intentional structure of roles for people to fill in an organization. It is intentional in the sense of making sure that all tasks necessary to accomplish goals are assigned and, it is hoped, assigned to people who can best accomplish them. Imagine what would have happened if such assignments had not been made during the special aircraft Voyager's flight around the globe without stopping or refueling. The purpose of an organizational structure is to help in creating an environment for human performance. It is, then, a management tool and not an end in and of itself. Although the structure must define the tasks to be done, the roles established must also be designed in light of the workers' abilities and motivations.

1.2.3 Staffing. *Staffing* involves filling, and keeping filled, the positions in the organizational structure. This is done by identifying workforce requirements, inventorying the people available, recruiting, selecting, placing, promoting, planning the career, compensating, and training or otherwise developing both candidates and current job holders to accomplish their tasks effectively and efficiently.

1.2.4 Leading. *Leading* (a.k.a. *directing*) is influencing people so they will contribute to organizational and group goals; it has to do predominantly with the interpersonal aspect of managing. All managers would agree that their most important problems arise from people—their desires and attitudes, their behavior as individuals and in groups—and those effective managers also need to be effective leaders.

Since leadership implies followership and people tend to follow those who offer a means of satisfying their own needs, wishes, and desires, it is understandable that leading involves motivation, leadership styles, and approaches and communication.

1.2.5 Controlling. *Controlling* is the measuring and correcting of activities of subordinates to ensure that events conform to plans. It measures performance against goals and plans, shows where negative deviations exist and, by putting in motion actions to correct deviations, helps ensure accomplishment of plans. Although planning must precede controlling, plans are not self-achieving. The plan guides managers in the use of resources to accomplish specific goals. Activities are then checked to determine whether they conform to plans.

Control activities generally relate to the measurement of achievement. Some means of controlling, like a budget for expense, inspection records, and the record of labor hours lost, are generally familiar. Each measures and shows whether plans are working. If deviations persist, correction is indicated, but what is corrected?

Nothing can be done about reducing scrap, for example, or buying according to specifications, or handling sales returns unless one knows who is responsible for these functions. Compelling events require locating the persons who are responsible for results that differ from planned action, and then taking the necessary steps to improve performance.

1.3 Management at different organizational levels.

All managers carry out managerial functions. However, the time spent for each function may differ. Figure 1 shows an approximation of the relative time spent for each function. Thus, top-level managers spend more time on planning and organizing than do lower-level managers. Leading, on the other hand, takes a great deal of time for first-line supervisors. The difference in time spent on controlling varies only slightly for managers at various levels.

1.4 Managerial skills.

Robert L. Katz [1955] and [1974] identified three kinds of skills for administrators—technical, human, and conceptual. To these may be added a fourth—the ability to design solutions.

(1) *Technical skill* is knowledge of and proficiency in activities involving

methods, processes, and procedures. Thus, it involves working with tools and specific techniques. For example, mechanics work with tools and their supervisors should have the ability to teach them how to use these tools.

(2) *Human skill* is the ability to work with people; it is cooperative effort; it is teamwork; it is the creation of an environment where people feel secure and free to express their opinions.

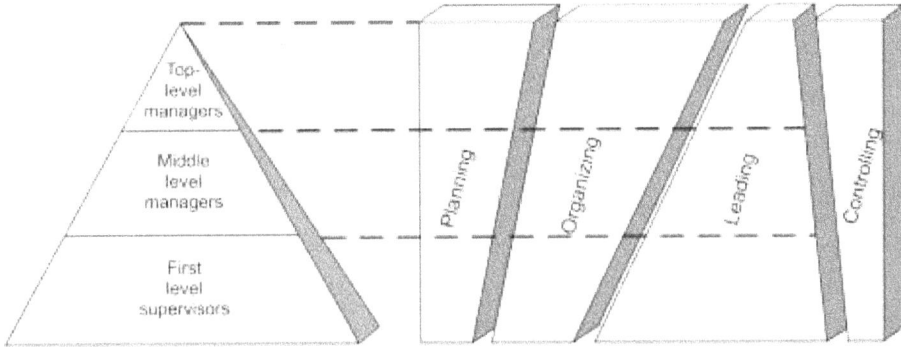

Figure 1: Time spent in carrying out managerial functions

(3) *Conceptual skill* is the ability to see the "big picture," to recognize significant elements in a situation, and to understand the relationships among the elements.

(4) *Design skill* is the ability to solve problems in ways that will benefit the enterprise. To be effective, particularly at upper organizational levels, managers must be able to do more than see a problem. If managers merely see the problem and become "problem watchers," they will fail. They must have, in addition, the skill of a good design engineer when working out a practical solution to a problem.

The relative importance of these skills may differ at various levels in the organizational hierarchy. As shown in Figure 2, technical skills are of greatest importance at the supervisory level. Human skills are also helpful during the frequent interactions with subordinates. Conceptual skills, on the other hand, are usually not critical for lower level supervisors. At the middle-management level, the need for technical skills decreases; human skills are still essential; conceptual skills gain in importance. At the top-management level, conceptual and design abilities and human skills are especially valuable, but there is relatively little need for technical abilities.

Tools that aid in the control of a SWE project are PERT and CPM, workload charts, and Gantt charts. The paper authored by Howes [1984] presents the *earned-value* method of tracking a SWE project.

The project manager is responsible for establishing the methods of monitoring the software project and reporting project status. Monitoring and reporting systems must be specified in order to determine project status. The project manager needs feedback detailing the progress of the project and quality of the product to ensure that everything is proceeding according to plan.

Managers must establish an environment in which every staff member can accomplish group goals with the least amount of time, money, materials, and personal dissatisfaction, or where they can achieve as much as possible of a desired goal with available resources.

In a nonbusiness enterprise such as a police department, as well as in units of a business (such as an accounting department) that are not responsible for total business profits, managers still have budgetary and organizational goals and should strive to accomplish them with a minimum of resources. The type, frequency, originator, and recipient of project reports must be specified. Status reporting tools to provide visibility of progress and not just resources used or time elapsed must be implemented. Tools that aid in the control and tracking of a SWE project are PERT and CPM, workload charts, Gantt charts, and the *earned-value method* Howes [1984].

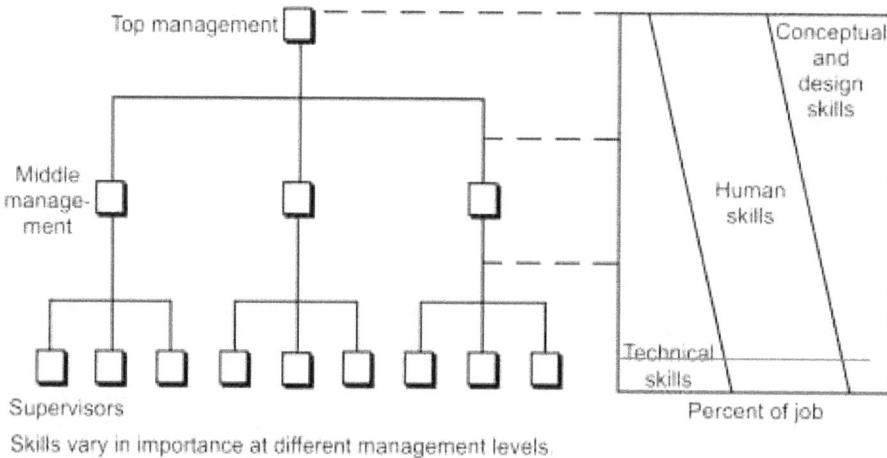

Skills vary in importance at different management levels.

Figure 2: Skills versus management levels

1.5 Productivity, effectiveness, and efficiency.

Another way to view the aim of all managers is to say that they must be productive. After World War II, the United States was the world leader in productivity. However, in the late 1960s, American productivity began to decrease at an accelerate rate.

Today government, private industry, and universities recognize the urgent need for productivity improvement. Until very recently, we frequently looked to Japan to find answers to our productivity problem, but this tendency overlooks the importance of effectively performing fundamental managerial and non-managerial activities.

1.5.1 Definition of productivity. Successful companies create a surplus through productive operations. Although the meaning of productivity varies between companies, we will define it as the *output-input ratio within a period with due consideration for quality.*

It can also be expressed as:

Productivity = output/input (within a time period—quality considered)

Thus, productivity can be improved by increasing outputs with the same inputs, by decreasing inputs but maintaining the same outputs, or by increasing output and decreasing inputs to change the ratio favorably. In the past, productivity improvement programs were mostly aimed at the worker level. Yet, as Peter F. Drucker, one of the most prolific writers in management observed, "The greatest opportunity for increasing productivity is surely to be found in knowledge, work itself and especially in management."

1.5.2 Definitions of effectiveness and efficiency. Productivity implies effectiveness and efficiency in individual and organizational performance. Effectiveness is the achievement of objectives. Efficiency is the achievement of the ends with the least amount of resources. To know whether they are productive, managers must know their goals and those of the organization.

2. Managing: Science or Art?

Managing, like so many other disciplines—medicine, music composition, engineering, accountancy, or even baseball—is in large measure an art but founded on a wealth of science. It is making decisions based on business realities. Yet, by applying decisions based on business realities, managers can in turn apply accrued organizational management knowledge to their respective projects. It is this knowledge, whether crude or advanced, whether exact or inexact, that, to the extent it is well organized, clear and pertinent, constitutes a science. Thus, managing as practiced is an art; the organized knowledge underlying the practice may be referred to as a science. In this context, science and art are not mutually exclusive but are complementary.

As science improves so should the application of this science (the art) as has happened in the physical and biological sciences. This is true because the many variables with which managers deal are extremely complex and intangible. However, such management knowledge as is available can certainly improve managerial practice. Physicians without the advantage of science would be little more than witch doctors. Executives who attempt to manage without such management science must trust luck, intuition, or their experiences.

In managing, as in any other field, unless practitioners are to learn by trial and error (and it has been said that managers' errors are their subordinates' trials), there is no place they can turn for meaningful guidance other than the accumulated knowledge underlying their practice.

3. The Elements of Science

Science is organized knowledge. The essential feature of any science is the application of the scientific method to the development of knowledge. Thus, we speak of a science as having clear concepts, theories, and other accumulated knowledge developed from hypotheses (assumptions that something is true), experimentation and analysis.

3.1 The scientific approach.

The *scientific approach* first requires clear concepts—mental images of anything formed by generalization from particulars. These words and terms should be exact, relevant to the things being analyzed and informative to the scientist and practitioner alike. From this base, the scientific method involves determining facts through *observation*. After classifying and analyzing these facts, scientists look for causal relationships. Facts are tested for generalizations or hypotheses that appear to be true, that is, to reflect or explain reality, and therefore, to have value in predicting what will happen in similar circumstances. These facts are called *principles*. This designation does not always imply that they are unquestionably or invariably true, but that they are believed to be valid enough to be used for prediction.

Theory is a systematic grouping of interdependent concepts and principles that form a framework for a significant body of knowledge. Scattered data, such as what we may find on a blackboard after a group of engineers has been discussing a problem, do not constitute information unless the observer has knowledge of the theory that will explain relationships. Theory is, as C.G. Homans [1958] has said, "in its lowest form a classification, a set of pigeonholes, a filing cabinet in which fact can accumulate. Nothing is more lost than a loose fact."

3.2 The role of management theory.

In the field of management, then, the *role of theory* is to provide a means of classifying significant and pertinent management knowledge. In designing an effective organizational structure, for example, a number of principles are interrelated and have a predictive value for managers. Some principles give guidelines for delegating authority; these include the principle of delegating by results expected, the principle of equality of authority and responsibility, and the principle of unity of command.

Principles in management are *fundamental truths* (or what are thought to be truths at a given time), explaining relationships between two or more sets of variables, usually an independent variable and a dependent variable. Principles may be descriptive or predictive and are not prescriptive. That is, they describe how one variable relates to another—what will happen when these variables interact. They do not prescribe what we should do. For example, in physics, if gravity is the only force acting on a falling body, the body will fall at an increasing speed; this principle does not tell us whether anyone should jump off the roof of a high building.

On the other hand, take the example of *Parkinson's Law:* work tends to expand to fill the time available. Even if Parkinson's somewhat frivolous principle is correct (as it probably is), it does not mean that a manager should lengthen the time available for the staff to do its job.

To take another example, in management the principle of *unity of command* states that the more often an individual reports to a single superior, the more that individual is likely to feel a sense of loyalty and obligation and the less likely it is that there will be confusion about instruction. The principle merely predicts. This principle in no sense implies that individuals should never report to more than one person. Rather, it implies that if they do so, their managers must be aware of the possible dangers and should consider these risks when balancing the advantages and disadvantages of multiple commands.

Like engineers who apply physical principles to the dsign of an instrument, managers who apply theory to managing must usually blend principles with realities. A design engineer is often faced with the necessity of combining considerations of weight, size, conductivity, and other factors. Likewise, a manager may find that the advantages of giving a controller authority to prescribe accounting procedures throughout an organization outweigh the possible costs of multiple authorities.

Nevertheless, if they know theory, these managers will know that such costs as conflicting instructions and confusion may exist and they will take steps—such as making the controller's special authority clear to everyone involved—to minimize or outweigh any disadvantages.

3.3 Management techniques.

Techniques are essentially ways of doing things and methods of accomplishing a given result. In all fields of practice they are important. They certainly are in managing, even though few important managerial techniques have been invented. Among them are budgeting, cost accounting, network planning, and controlling techniques like the Program Evaluation and Review Technique (PERT) or the Critical Path Method (CPM), rate-of-return-on-investment control, various devices of organizational development, managing by objectives, and Total Quality Management (TQM).

4. Systems Approach to Operational Management

An *organized enterprise* does not exist in a vacuum. Rather, it depends on its external environment; it is a part of larger systems such as the industry to which it belongs, the economic system, and society. Thus, the enterprise receives inputs, transforms them, and exports the outputs to the environment. However, this simple model needs to be expanded and developed into a model of operational management that indicates how the various inputs are transformed through the managerial functions of *planning, organizing, staffing, leading,* and *controlling.* Clearly, any business or other organization must be described by an open-system model that includes interactions between the enterprise and its external environment.

4.1 Inputs and stakeholders.

The *inputs* from the external environment may include people, capital, and managerial skills, as well as technical knowledge and skills. In addition, various groups of people make demands on the enterprise. For example, employees want higher pay, more benefits, and job security. On the other hand, consumers demand safe and reliable products at reasonable prices. Suppliers want assurance that their products will be bought.

Stockholders want not only a high return on their investment but also security for their money. Federal, state, and local governments depend on taxes paid by the enterprise, but they also expect the enterprise to comply with their laws. Similarly, the community demands that enterprises be "good citizens," providing the maximum number of jobs with a minimum amount of pollution.

Other claimants to the enterprise may include financial institutions and labor unions; even competitors have a legitimate claim for fair play. It is clear that many of these claims are incongruent and it is the manager's job to integrate the legitimate objectives of the claimants.

4.2 The managerial transformation process.

Managers have the task of transforming inputs, effectively and efficiently, into outputs. Of course, the transformation process can be viewed from different perspectives. Thus, one can focus on such diverse enterprise functions as finance, production, personnel, and marketing. Writers about management look at the transformation process in terms of their particular approaches to management.

Specifically, as you will see, writers belonging to the human behavior school focus on interpersonal relationships; social systems theorists analyze the transformation by concentrating on social interactions; and those advocating decision theory see the transformation as sets of decisions. However, we believe that the most comprehensive and useful approach for discussing the job of managers is to use the managerial functions of *planning, organizing, staffing, leading,* and *controlling* as a framework for organizing managerial knowledge.

4.3 The communication system.

Communication is essential to all phases of the managerial process. It integrates the managerial functions and links the enterprise with its environment. A communication system is a set of information providers and information recipients, along with the means of transferring information from one group to another group, with the understanding that the messages being transmitted will be understood by both groups. For example, the objectives set in planning are communicated so that the appropriate organizational structure can be devised.

Communication is essential in the selection, appraisal, and training of managers to fill the roles in this structure. Similarly, effective leadership and the creation of an environment conducive to motivation depend on communication. Moreover, it is through communication that one determines whether events and performance conform to plans. Thus, communication makes managing possible.

The second function of the communication system is to link the enterprise with its external environment, where many of the stockholders are located. Effective managers will regularly scan the external environment. While it is true that managers may have little or no power to change the external environment, they have no alternative but to respond to it. For example, one should never forget that the customer, who is the reason for the existence of virtually all businesses, is situated outside a company.

It is through the communication system that the needs of customers are identified; this knowledge enables the firm to provide products and services at a profit. Similarly, it is through an effective communication system that the organization becomes aware of competition and other potential threats and constraining factors.

4.4 Outputs.

Managers must secure and utilize inputs to the enterprise to transform them through the managerial functions—with due consideration for external variables—to produce outputs.

Although the kinds of outputs will vary with the enterprise, they usually include a combination of products, services, profits, satisfaction, and integration of the goals of various claimants to the enterprise. Most of these outputs require no elaboration and only the last two will be discussed.

The organization must indeed provide many "satisfactions" if it hopes to retain and elicit contributions from its members. It must contribute to the satisfaction not only of basic material needs (for example, earning money to buy food and shelter or having job security) but also of needs for affiliation, acceptance, esteem, and perhaps even self-actualization.

Another output is goal integration. As noted above, the different claimants to the enterprise have very divergent—and often directly opposing—objectives. It is the task of managers to resolve conflicts and integrate these aims. This is not easy, as one former Volkswagen executive discovered. Economics dictated the construction of a Volkswagen assembly plant in the United States. However, an important claimant, German labor, out of fear that jobs would be eliminated in Germany, opposed this plan. This example illustrates the importance of integrating the goals of various claimants to the enterprise, which is indeed an essential task of any manager.

4.5 Providing feedback to the system.

Finally, we should notice that in the systems model of operational management, some of the outputs become inputs again. Thus, the satisfaction of employees becomes an important human input to the enterprise. Similarly, profits, the surplus of income over costs, are reinvested in cash and capital goods, such as machinery, equipment, buildings, and inventory.

4.6 Coordination, the essence of managership.

Some authorities consider coordination to be an additional function of management. It seems more accurate, however, to regard it as the essence of managership, where the management purpose is to harmonize individual efforts in the accomplishment of group goals. Each of the managerial functions is an exercise contributing to coordination.

Even in the case of a church or a fraternal organization, individuals often interpret similar interests in different ways and their efforts toward mutual goals do not automatically mesh with the efforts of others. It thus becomes the central task of the manager to reconcile differences in approach, timing, effort, or interest, and to harmonize individual goals to contribute to organizational goals.

5. Summary

Management is the process of designing and maintaining an environment in which individuals, working together in groups, accomplish efficiently selected aims. Managers are charged with the responsibility of taking actions that will make it possible for individuals to make their best contributions to group objectives. Managing as practiced is an art; the organized knowledge underlying the practice may be referred to as a science. In this context, science and art are not mutually exclusive but are complementary.

REFERENCES

- **[Homans 1958]** C.G. Homans, "Social Behavior as Exchange," *American Journal of Sociology,* 63:597-606, 1958.

- **[Howes 1984]** N.R. Howes, "Managing Software Development Projects for Maximum Productivity," *IEEE Transactions on Software Engineering*, Volume: SE-10, Issue: 1, Jan. 1984,

- **[Katz 1955]** Robert L. Katz, "Skills of an Effective Administrator," *Harvard Business Review*, Jan-Feb, 1955, pp. 33-42.

- **[Katz 1974]** Robert L., Katz, "Retrospective Commentary," *Harvard Business Review*, Sep-Oct, 1974, pp. 101-102.

Chapter 4

A Standard for Software Project Management Plans[4]

Abstract

Planning a software engineering project consists of management activities selected to determine 1) future courses of action for the project, and 2) a process for completing those actions. "Planning is deciding in advance what to do, how to do it, when to do it, and who is to do it" [Koontz, O'Donnell and Weihrich 1980].

The process of developing software engineering project plans is outlined. The format and contents of software project management plans, applicable to any type or size of software project, are described. The elements that should appear in all software project management plans are identified.

1. Introduction

1.1 Purpose.

This standard specifies the method of development and the format and content of software project management plans. This standard does not specify the exact techniques to be used in developing a software project management plan, nor does it provide examples of software project management plans. Each organization using this standard should develop a set of practices and procedures to provide detailed guidance for preparing and updating software project management plans based on this standard. These practices and procedures should take into account the environmental, organizational, and political factors that influence application of the standard.

Not all software projects are concerned with development of source code for a new software product. Some software projects consist of a feasibility study and definition of product requirements. Other software projects terminate upon completion of product design, and some projects are concerned with major modifications to existing software products. This standard is applicable to all

---

4. Chapter 4 is modified *from IEEE Standard for Software Project Management Plans*. This standard has been developed for a software engineering classroom. This classroom standard should not be used to satisfy a commercial software engineering contract. Nevertheless, it can be used as (1) an educational tool and (2) a classroom standard for students to use when preparing a classroom software project management specification.

types of software projects; applicability is not limited to projects that develop source code for new products. Project size or type of software product does not limit application of this standard. Small projects may require less formality in planning than large projects, but all components of the standard should be addressed by every software project.

Software projects are sometimes component parts of larger projects. In these cases, the software project management plan may be a separate component of a larger plan or it may be merged into a system-level or business-level project management plan.

> *Remember to base your plans on the job to be done—the software requirements specifications.*

In this subsection prepare a brief statement of the business or system needs to be satisfied by the project, including a concise summary of the project objectives, the products to be delivered to satisfy those objectives, and the methods by which satisfaction will be determined. The project statement of purpose describes the relationship of this project to other projects, and as appropriate, how this project will be integrated with other projects or ongoing work processes.

1.2 Scope of the planning standard.

The subsection should:

(1) Identify the software product(s) to be produced by name. For example; Host DBMS, Report Generator, AJAX Parole Program, F-1 Fire Control System, etc.

(2) Explain what the software product(s) will, and, if necessary, will not do.

(3) Describe the application of the software being specified. A portion of this application should:

    a. Describe all relevant benefits, objectives, and goals as precisely as possible. For example, to say that one goal is to provide effective reporting capabilities is not as effective as saying parameter-driven, user-definable reports with a two-hour turnaround and on-line entry of user parameters.

    b. Be consistent with similar statements in higher-level specifications (system/software requirements specifications, software contract, statement of work), if they exist.

1.3 Definitions, acronyms, and abbreviations.

This subsection should provide the definitions of all terms, acronyms, and abbreviations required to properly interpret the SDD. This information may be proven by reference to one or more appendices in the SDD or by reference to other documents.

1.4 References.

This subsection should provide a list of cited references and a separate list of supporting references. Each reference should:

(1) Provide a complete list of all documents referenced elsewhere in the SDD, or in a separate, specified document.

(2) Identify each document by title; report number, if applicable; date; and publishing organization.

(3) Specify the sources from which the references can be obtained. This information may be provided by reference to an appendix or to another document.

1.5 Assumptions and constraints.

The project management plans describe the assumptions upon which the project is based and imposed constraints on project factors such as schedule, budget, resources, software to be reused, acquired software to be incorporated, technology to be employed, and product interfaces with other products.

1.6 Audience.

This standard is intended for use by software project managers and other personnel who prepare and update project plans and monitor adherence to those plans.

1.7 Evolution of Plans.

Developing the initial version of the software project management plan should be one of the first activities to be completed in a software project. As the project evolves, the nature of the work to be done will be better understood and plans will become more detailed. Thus, each version of the plan should be placed under configuration management, and each version should contain a schedule for subsequent updates to the plan.

1.8 Terminology.

The word *shall* is used to indicate mandatory requirements contractually binding, meaning the requirements must be implemented, and their implementation verified.

The word *should* is used to indicate that among several possibilities one is recommended as particularly suitable, without mentioning or excluding others; or that a certain course of action is preferred but not necessarily required; or that (in the negative form) a certain course of action is deprecated but not prohibited.

The word *may* is used to indicate a course of action permissible within the limits of the standard.

Notes can also be used to provide auxiliary information to the software development organization. These are <u>not</u> classically considered to be "contractually binding" requirements.

1.9 Overview.

This guide presents the software project management process and the format

and contents of a software engineering project-planning document to be used when developing or modifying a software-intensive system.

Planning a software engineering project consists of management activities selected to determine 1) future courses of action for the project, and 2) a process for completing those actions. "Planning is deciding in advance what to do, how to do it, when to do, and who is to do it" [5].

Each software engineering project should begin with a workable plan. Uncertainties and unknowns make planning necessary both within the software project environment and with external sources. Planning focuses attention on project goals, actions necessary to reach those goals, and potential risks and problems that might interfere with obtaining those goals.

The software project management plan is the controlling document for managing a software project. The project management plan is one of the two foundation documents for a software engineering project (the software requirements specifications being the other). A software project management plan defines the technical and managerial functions, activities, and tasks necessary to satisfy the requirements of a software project, as defined in the project agreement.

The plan provides an advance notice to all concerned regarding how the project manager (PM) intends to manage the project. It also serves as a "letter of agreement" with supporting agencies detailing what kind of support is to be given to the project, how much is to be given, when it is to be given, and the cost of this service (if any).

This standard describes the necessary structure, contents, and steps to be taken to design a complete and effective software project management plan. Section 3 of this chapter provides an annotated outline for this project plan. (*See plans Section 3, p 88*).

2. Start-up Project Management Plan

Software project management is a system of procedures, practices, technologies, and expertise that provides the planning, organizing, staffing, directing, and controlling necessary to manage successfully a software engineering project.

---

5. From *Management*, Koontz and O'Donnell, 1972.

(Know-how in this case means the expertise, skill, background, and wisdom to effectively apply knowledge in practice).

One of the project manager's primary duties is to assure that a correct and complete project management plan is developed and maintained throughout the project's life cycle.

Planning a software engineering project consists of the management activities that lead to selecting, among alternatives, future courses of action for the project and a program for completing those actions.

There is a substantial amount of work and effort involved with starting a software engineering project. Therefore, it is important that the start-up phase be planned and scheduled with a tracked expenditure.

2.1 Initial cadre.

The first step in the planning process is to assemble a small group of software engineering developers to design the initial planning for the start-up effort. This initial group can become the core of the project staff. Although the size of this initial group depends on the size of the finished product, it is typically composed of three to seven very qualified developers with a designated group leader. A senior and experienced manager would lead this initial cadre of developers and provide the primary interaction between the customer and senior management to ensure that contractual requirements are met.

2.2 Start-up plan.

The cadre is responsible for planning the start-up effort. This includes:

(1) Scoping the work to be done.

(2) Identifying the target environment.

(3) Determining what personnel, materials, technology, and environments are available, adequate, and appropriate to manage the project.

(4) Developing long-term achievable time-scales to completion.

(5) Identifying long-term costs needed to support the product.

(6) Developing a start-up plan, including cost and schedule for the initial planning, organizing, staffing, and control of the project.

(7) Reporting the start-up plan to the customer and senior management, and obtaining permission to proceed.

3. Managerial Process Plan

This clause of the SEPM specifies the project management processes. This clause should be written in a manner consistent with the statement of project scope, including the project start-up plan, the estimation plans, the project work plan, the project control plan, the risk management plan, and the project closeout plan.

3.1 Estimation plan.

Specify the estimation time and estimation techniques used to determine cost, schedule, staffing requirements, and staff-training requirements for the complete project.

3.1.1 Cost and schedule plan. Detail the cost and schedule for conducting the project as well as methods, tools, and techniques used to estimate project cost, schedule, resource requirements, and associated confidence levels. In addition, specify the basis of estimation, including techniques such as analogy, rule of thumb, local history, and the sources of data. This subsection should also specify the methods, tools, and techniques that will be used to periodically re-estimate the cost, schedule, and resources needed to complete the project. Periodically schedule re-estimation (perhaps monthly) as necessary.

3.1.2 Staffing plan. The number of staff required for each skill level, the numbers of personnel and types of skills necessary for each project phase, and the duration of need must be specified in advance. In addition, this subsection shall specify the sources of staff personnel; for example by internal transfer, new hire, or contract. Resource Gantt charts, resource histograms, spreadsheets, and tables may be used to depict the staffing plan by skill level, by project phase, and by aggregations of skill levels and project phases.

3.1.3 Resource acquisition plan. Devise a plan for the acquisition of non-personnel resources necessary to successfully complete the project. The resource acquisition plan should include a description of the resource acquisition process, including assignment of responsibility to all aspects of resource acquisition. The plan should include, but not be limited to, acquisition plans for equipment, computer hardware and software, training, service contracts, transportation, facilities, and administrative and janitorial services. The plan should specify when each acquisition activity will be required. Specify constraints pertaining to acquiring the necessary resources. This subsection may be expanded into additional subsections to accommodate acquisition plans for various types of resources to be acquired.

3.1.4 Project-staff training plan. Training is required to ensure that sufficient numbers of necessary skill levels are available to successfully design the software project. The training schedule should include the types of training to be provided, numbers of personnel to be trained, entry and exit criteria for training, and the training method; for example, lectures, consultations, mentoring, or computer assisted training. Offer training sessions in both technical and managerial skills as needed.

The three most important parts of a project plan are *cost, schedule,* and *tasks to be performed*

3.2 Work Plan.

The work plan outlines the work activities, schedule, resources, and budget details for the software project.

3.2.1 Work activities. The various work activities to be performed in the software project must be analyzed ahead of time. A work breakdown structure is the best method used to depict the work activities and their relationships to one another. Work activities should be decomposed to a level that exposes all project risk factors and allows accurate estimates of resource requirements and schedule duration for each work activity.

Work packages should be used to specify, for each work activity, factors such as necessary resources, estimated duration, work products to be produced, acceptance criteria for the work products, and predecessor and successor work activities. The level of decomposition for different work activities in the work breakdown structure may differ, depending on factors such as when the work package is scheduled to be accomplished, the quality of the requirements, familiarity of the work, and novelty of the technology employed.

3.2.2 Schedule allocation. Scheduling relationships among work activities is required. Scheduling should be prioritized in a manner that depicts time-sequencing constraints and illustrates opportunities for concurrent work activities. Any scheduling constraints on particular work activities caused by factors external to the project are to be indicated in the work activity schedule. The schedule should include frequent milestones that can be assessed for achievement using objective indicators to assess the scope and quality of work products completed at those milestones. Techniques for depicting schedule relationships may include milestone charts, activity lists, activity Gantt charts, activity networks, critical path networks (CPN), and PERT.

3.3 Resource allocation.

Prepare a detailed itemization of the resources allocated to each major work activity in the project work breakdown structure. Resources include the numbers and required skill levels of personnel for each work activity. Allocation may include, as appropriate, personnel by skill level and factors such as computing resources, software tools, special testing and simulation facilities, and administrative support. Provide a separate line item for each type of resource within each work activity. Collect a summary of resource requirements for the various work activities from the work packages of the work breakdown structure and present them in tabular form.

3.3.1 Budget allocation. Prepare a detailed breakdown of necessary resource budgets for each of the major work activities in the work breakdown structure. The activity budget should detail the estimated cost for activity personnel and may include, as appropriate, costs for factors such as travel, meetings, computing resources, software tools, special testing and simulation facilities, and administrative support. Use a separate line itemization for each type of resource within each activity budget. The work activity budget may be developed using a spreadsheet and presented in tabular form.

3.4 Control plan.

Specify the metrics, reporting mechanisms, and control procedures necessary to measure, report, and control the status of product requirements, the project schedule, the budget, and the quality of work processes and work products. All elements of the control plan should be consistent with the organization's standards, policies, and procedures for controlling software projects, and with any contractual agreements for project control.

3.4.1 Requirements control plan. Outline the control mechanisms for measuring, reporting, and controlling changes to the product requirements. Specify the mechanisms to be used in assessing the impact of requirements changes on product scope and quality and the impacts of requirements changes on project schedule, budget, resources, and risk factors. Configuration management mechanisms should include change control procedures and a change control board. Techniques that may be used for requirements control include traceability, prototyping and modeling, impact analysis, and reviews.

3.4.2 Schedule control plan. Determine the control mechanisms to be used to measure the progress of work completed at the major and minor project milestones, to compare actual progress to planned progress, and to implement corrective action when actual progress does not conform to planned progress. This plan specifies the methods and tools used to measure and control schedule progress. Achievement of schedule milestones should be assessed using objective criteria to measure the scope and quality of work products completed at each milestone.

3.4.3 Budget control plan. Decide which control mechanisms will be used to measure the cost of work completed, compare planned cost to budgeted cost, and implement corrective action when actual cost does not conform to budgeted cost. The budget control plan specifies at which intervals cost reporting will be scheduled and the methods and tools used to manage the budget. The budget plan should include frequent milestones that can be assessed for achievement using objective indicators to assess the scope and quality of work products completed at those milestones. Use a mechanism such as earned value tracking to report the budget and schedule plan, schedule progress, and calculate the cost of work completed.

Corrective action must be taken if the project is not proceeding according to plan.

3.4.4 Quality control plan. The mechanisms to be used to measure and control the quality of the work processes and the resulting work products must be specified. Quality control mechanisms may include quality assurance of work processes, verification and validation, joint reviews, audits, process assessment, and product evaluations.

3.4.5 Reporting plan. The reporting mechanisms, report formats, and information flows to be used in communicating the status of requirements, schedule, budget, quality, and other desired or required status metrics must be specified. Specify each, both within the project and as it applies to entities external to the project. Include the methods, tools, and techniques of communication in this subsection. The frequency and detail of communications related to project measurement and control should be consistent with the project scope, criticality, risk, and visibility.

3.4.6 Metrics plan. Outline the methods, tools, and techniques to be used in collecting and retaining project metrics. The metrics collection plan specifies the metrics to be collected, the frequency of collection, and the methods to be used in validating, analyzing, and reporting the metrics. In addition, the metrics plan provides the rationale for the particular metrics chosen for the project.

3.5 Risk management plan.

This subsection of the SPMP shall specify a risk management plan to identify, analyze, and prioritize project risk factors. Describe the procedures for contingency planning, and the methods to be used in tracking the various risk factors, evaluating changes in the levels of risk factors, and the responses to those changes. The risk management plan shall also specify plans for assessing initial risk factors, and the on-going identification, assessment, and mitigation of risk factors throughout the project life cycle.

When completed, this plan should describe risk management work activities, procedures, and schedules for performing those activities, documentation, and reporting requirements, organizations, personnel responsible for performing specific activities, and procedures for communicating risks, and risk status among the various acquirer, supplier, and subcontractor organizations. Risk factors include risks in the acquirer-supplier relationship, contractual risks, technological risks, risks caused by the size and complexity of the product, risks in the development and target environments, risks in personnel acquisition, skill levels, and retention, risks to schedule and budget, and risks in achieving acquirer acceptance of the product.

3.6 Project closeout plan.

The project closeout plan contains the plans necessary to ensure orderly closeout of the software project. Individual items in the closeout plan include a staff reassignment plan, a plan for archiving project materials, a plan for postmortem debriefings of project personnel, preparation of a final report to include lessons learned, and an analysis of project objectives achieved.

4. Technical Process Plan

The technical process plan clause contains the following: the development process model, the technical methods, tools, and techniques to be used to develop the various work products, plans for establishing and maintaining the project infrastructure, and the product acceptance plan.

4.1 Process model.

The relationships among major project work activities and supporting processes are defined in the process model. Definitions are achieved by determining the flow of information and work products among activities and functions, the timing of work products to be generated, reviews to be conducted, major milestones to be achieved, baselines to be established, project deliverables to be completed, and required approvals that span the duration of the project. The process model for the project includes project initiation and project termination activities. A combination of graphical and textual notations may be used to describe the process model. If an organization's standard process model must be tailored to a project, indicate this in the process model section.

4.2 Methods, tools, and techniques.

Determine the development methodologies, programming languages, and other notations, and the tools and techniques to be used to specify, design, build, test, integrate, document, deliver, modify, and maintain the project deliverable and non-deliverable work products. In addition, decide which technical standards, policies, and procedures governing development and/or modification of the work products will be used.

4.3 Infrastructure plan.

Devise a plan for establishing and maintaining the development environment (hardware, operating system, network, and software), and the policies, procedures, standards, and facilities required to conduct the software project. These resources may include workstations, local area networks, desks, office space, and provisions for physical security, administrative personnel, and janitorial services. Additional development resources include software tools for analysis, design, implementation, testing, and project management.

4.4 Product acceptance plan.

This subsection of the SEPM should prepare the plan to be used for acquirer acceptance of the deliverable work products generated by the software project. Clearly specify the objective criteria for determining acceptability of the deliverable work products included in this plan. Ensure that representatives of the development organization and the acquiring organization sign a formal agreement detailing the acceptance criteria. Include technical processes, methods, or tools required for product acceptance in the product acceptance plan. Methods such as testing, demonstration, analysis, and inspection should also be specified in this plan.

5. Supporting Process Plan

This clause of the SEPM shall contain plans for the supporting processes that span the duration of the software project. These plans include, but are not limited to, configuration management, verification and validation, software documentation, quality assurance, reviews and audits, problem resolution, and subcontractor management. Plans for supporting processes are developed to a level of detail consistent with the other clauses and subsections of the SEPM. In particular, the roles, responsibilities, authorities, schedule, budgets, resource requirements, risk factors, and work products for each supporting process are specified.

The nature and types of required supporting processes may vary from project to project. Explicitly justify the absence of a configuration management plan, verification and validation plan, documentation plan, quality assurance plan, joint acquirer-supplier review plan, problem resolution plan, or subcontractor management plan in any software project management plan in which one or more of these plans is not included. Plans for supporting processes may be either incorporated directly into the software project management plan or incorporated by reference to other plans.

5.1 Configuration management plan.

Prepare a management plan for the software project. Include the methods that will be used to provide configuration identification, control, status accounting, evaluation, and release management. In addition, specify the processes of configuration management to include procedures for initial baselining of work products, logging, and analysis of change requests, change control board procedures, tracking of changes in progress, and procedures for notifying concerned parties when baselines are first established or later changed. The configuration management process should be supported by one or more automated configuration management tools.

5.2 Verification and validation plan.

Develop the verification and validation plan for the software project. Include scope, tools, techniques, and responsibilities for the verification and validation work activities. Outline the organizational relationships and degrees of independence between development activities and verification and validation activities.

Verification planning results in specification of techniques such as traceability, milestone reviews, progress reviews, peer reviews, prototyping, simulation, and modeling.

Validation planning results in specification of techniques such as testing, demonstration, analysis, and inspection. Lastly, specify automated tools used in verification and validation.

5.3 Documentation plan.

A documentation plan is required for the software project. Plans for generating

non-deliverable and deliverable work products are included. Organizational entities responsible for providing input information, and generating and reviewing the various documents are specified in this plan. Non-deliverable work products may include items such as requirements specifications, design documentation, traceability matrices, test plans, meeting minutes, and review reports.

Deliverable work products may include source code, object code, users' manual, on-line help system, regression test suite, configuration library, principles of operation, a maintenance guide, or other items specified in the software project management plan. The final documentation plan should include a list of documents to be prepared, the controlling template or standard for each document, who will prepare it, who will review it, due dates for the review copy and initial baseline version, and a distribution list for review copies and baseline versions.

5.4 Quality assurance plan.

The quality assurance plan ensures that: (1) the software project fulfills its commitments to the software process, and (2) the software product is specified in the requirements specification, the software project management plan, supporting plans, and any standards, procedures, or guidelines to which the process or the product must adhere. Quality assurance procedures may include analysis, inspections, reviews, audits, and assessments. The quality assurance plan should indicate the relationships among the quality assurance, verification and validation, review, audit, configuration management, system engineering, and assessment processes.

5.5 Reviews and audits plan.

Detail the schedule, resources, methods, and procedures to be used in conducting project reviews and audits. The review and audit plan should offer a blueprint for joint acquirer-supplier reviews, management progress reviews, developer peer reviews, quality assurance audits, and acquirer-conducted reviews and audits.

5.6 Problem resolution plan.

Determine the resources, methods, tools, techniques, and procedures to be used in reporting, analyzing, prioritizing, and processing software problem reports generated during the project. The problem resolution plan indicates the roles of development, configuration management, the change control board, and verification and validation in problem resolution work activities. Effort devoted to problem reporting, analysis, and resolution should be separately reported so that rework can be easily tracked for process improvement accomplishment.

5.7 Subcontractor management plans.

Develop the necessary plans for selecting and managing any subcontractors that may contribute work products to the software project. Define the criteria for selecting subcontractors in the management plan for each subcontract generated using a tailored version of this standard. Tailored plans should include the items necessary to ensure successful completion of each subcontract. In particular, requirements management, monitoring of technical progress, schedule and budget control, product acceptance criteria, and risk management procedures are included in each subcontractor plan. Additional topics are added as needed to ensure successful completion of the subcontract. Prepare a reference listing of the official subcontract and prime contractor/subcontractor points of contact.

5.8 Process improvement plan.

Devise plans to periodically assess the project, determine areas for improvement, and implement improvement plans. The process improvement plan should be closely related to the problem resolution plan; for example, root cause analysis of recurring problems may lead to simple process improvements that can significantly reduce rework during the remainder of the project. Examine implementation of improvement plans to identify those processes that can be improved without serious disruptions to an on-going project and to identify those processes that are easily improved by process improvement initiatives at the organizational level.

6. Additional Plans

Additional plans are required to satisfy product requirements and contractual terms. Additional plans for a particular project may include plans for assuring that safety, privacy, and security requirements for the product are met. They may also include special facilities or equipment, product installation plans, user-training plans, integration plans, data conversion plans, system transition plans, product maintenance plans, or product support plans. Devise additional project plans as necessary to meet these needs and other, unexpected needs that may arise.

Remember: A plan in the mind of man is not a plan at all, i.e., to be a real plan in must be documented.

Appendix.

Appendices may be included, either directly or by reference to other documents, to provide supporting details that could potentially detract from the SEPM if included in the body of the SEPM.

REFERENCE

- **[Koontz, O'Donnell and Weihrich 1980]** H. Koontz, C. O'Donnell and H. Weihrich, *Management*, 7th ed., McGraw-Hill Book Co., New York, 1980.

Chapter 5
Software Management Exercises

These exercises are provided to encourage you to browse the chapter looking for answers to the questions provided. If truth were told, the correct answer for all software engineering questions is "it depends." To avoid this issue, a set of possible answers are provided. There is (supposedly) only one correct answer.

If you are using this book as a textbook in a university course, your instructor may require you to justify your answer. (Why are some of the possible answers correct and why are some of them wrong?) The instructor might also ask you to identify any assumptions you depended on in arriving at your answer.

However, if you are very clever, maybe you can come up with more than one correct answer (which of course you have to justify).

1.   **Which of the following illustrates the software engineering indicators of a software project failure:**

   I.   **The errors in the software system caused a large amount of negative publicity for the development organization.**
   II. **The project did not meet its requirements.**
   III. **The project cost more than the customer expected.**
   IV. **The project manager was caught passing "trade secrets" to a competitor and was fired.**
   V.   **The delivered system was not a marketing success. The project was delivered six months late.**

   [a]   I, II, IV & V
   [b]   I, IV & V
   [c]   II, III, IV & V
   [d]   III, IV & V

2.   **If you are assigned a project with a fixed requirements and schedule, what is the third parameter that must be flexible?**

   [a] Cost
   [b] Opportunity
   [c] Risk
   [d] Quality

3.   A work breakdown structure is:

       I.   A means of representing a product
       II.   A means of representing a process
       III. Required in developing top-down software costs
       IV.   Only used in embedded computer systems

       [a] I only
       [b] I & II
       [c] I, II & III
       [d] II & IV

4.   Which of the following statements regarding schedule tools are correct?

       [a] The Gantt chart can be used to display presidencies between tasks.
       [b] CPM chart can show minimum delivery time.
       [c] The PERT chart can show very accurate delivery times.
       [d] The activity network can show numerous delivery times.

5.   When planning incremental builds, which of the following should be considered by
       the project manager?

       I.   Can the requirements be completed prior to beginning the first build?
       II.   Can the software system be portioned into subsystems?
       III. Are corresponding regression tests available?
       IV.   Can the subsystems be tested independently?

       [a] I & II
       [b] I, II & III
       [c] II & III
       [d] I, III & IV

6.   The concept of universality of management implies which of the follow-
       ing?

       [a] All managers perform the same functions.
       [b] All managers perform the same activities.
       [c] All managers perform the same tasks.
       [d] All managers are equal.

7.   Supervising personnel is an activity of which of the management functions
       below?

       [a] Organizing
       [b] Staffing
       [c] Directing
       [d] None of the above

8. According to Dr. Barry Boehm in his book *Software Engineering Economics,* which of the following statements is true?

    I. Personnel capabilities are the greatest cost drivers.
    II. The type of programming language used has a greater effect on software cost than any other cost attribute.
    III. Size of the computer program (in lines of code) does not affect the cost of building the system.
    V. Personnel turnover affects the cost of software development.
    VI. Project schedules that have been compressed by more than 25% cannot be delivered.

    [a] I, II, IV & V
    [b] I, IV & V
    [c] I & II
    [d] III, IV & V

9. A software project is estimated to take a nominal 12 months with a four-person team. Project stakeholders desire the project to be finished as soon as possible and suggest using an eight-person team to achieve that desire. Given this charge, the stakeholders can expect the project to be completed in:

    [a] 9 months
    [b] 7 months
    [c] 6 months
    [d] 4 months

10. Technical policies for planning & controlling the operation, use, & development of software systems are the responsibility of which of the following type of management?

    [a] Software acquisition management
    [b] Senior level management
    [c] Developer management
    [d] Risk management

11. A completed software product delivered for system integration corresponds to which of the following?

    [a] A functional baseline
    [b] An allocated baseline
    [c] A developmental baseline
    [d] A product baseline

12. Depending on the circumstances, software may be developed by a tempo-rary or a permanent organization. Temporary organizations are associat-ed with which set of the following development formats?

   I. Project format
   II. Matrix project format
   III. Functional project format

   [a] I & II
   [b[II & III
   [c[I & III
   [d] All of the above

13. Which two of the following statements concerning the critical path method used in project planning are true?

   I. The critical path method is an example of a manual method.
   II. The critical path method is an example of an automatic method.
   III. The critical path method is an example of a management method.
   IV. The critical path method is an example of a development method.

   [a] I & III
   [b] II & III
   [c] I & IV
   [d] II & IV

14. Which two of the following statements concerning the critical path method used in project planning are true?

   I. The critical path method is an example of a manual method
   II. The critical path method is an example of an automatic method
   III. The critical path method is an example of a management method
   IV. The critical path method is an example of a development method

   [a] I & III
   [b] II & III
   [c] I & IV
   [d] II & IV

15. Which of the following is NOT a valid use of a baseline?

[a] To distinguish between different internal releases for delivery to a customer
[b] To help ensure complete and up-to-date documentation
[c] To enforce standards
[d] To control changes to the executable code modules

16. A software project is estimated to take a nominal twelve months with a four-person team. Project stakeholders desire the project to be finished as soon as possible and suggest using an eight-person team to achieve that desire. Given this charge, the stakeholders can expect the project to be completed in:

[a] 9 months
[b] 7months
[c] 6 months
[d] 4 months

17. When a project deviates from plans or requirements the best possible action to take is which of the following?

[a] Initiate corrective action to bring the project into conformance with plans
[b] Change the plans to make them conform to the actual state of the project
[c] Cancel the project
[d] Do nothing

18. When planning incremental builds, which of the following should be considered by project manager?

I. Can the requirements be completed prior to beginning the first build?
II. Can the software system be portioned into subsystems?
III. Are there corresponding regression tests available?
IV. Can the subsystems be tested independently?

19.   The concept of universality of management implies which of the following?

[a] All managers perform the same functions.
[b] All managers perform the same activities.
[c] All managers perform the same tasks.
[d] All managers are equal.

20.   A work breakdown structure is:

I.   A means of representing a product
II. A means of representing a process
III. Required in developing top-down software costs
IV. Only used in embedded computer systems

[a] I only
[b] I and II
[c] I, II, and III
[d] II and IV

INDEX

document project plans, 10, 16
DoD Software Initiative, 8, 52, 57
Donnelly, James H. Jr., 4. 57
dual-career ladders, 66

earned-value, 45, 47, 75, 76
earned value tracking, 47
education, 29, 31
effectiveness, 32, 45, 52, 71, 77
efficiency, 40, 67, 71, 77
egoless programming teams (a.k.a. democratic team), 23
evaluate project personnel, 3, 32
experience, 27, 28, 29, 68
experience, amount of, 31
experience, diversity of, 31

facilitate communication, 3, 36, 41
Fayol, Henri, 4, 58
filling the positions, 29
forecast future events, 3, 10, 11
functional organizational structure, 20-21
functions of management, 72

Gantt charts, 13,47, 75, 76, 90, 91,100
general development of the project staff, 32
Gibson, James L., 4, 57
group affinity, 30

hierarchical teams, 23
Homan, C.G., 90
Howes, N.R., 56

independent auditing, 48, 49
inspections (*See also walkthroughs*), 48, 49
intelligence, 30
Ivancevich, John M., 4, 57

King, William R., 4, 57
Koontz, Harold, ix

leading, ix, 2, 33, 38, 71, 72, 73, 74, 80, 81
line organization, 19, 20

project management, 2
project management is management, v, 3, 4
project management methods, 13
project managers, 27, 28, 45
project plan, 3, 15, 16, 35, 64, 67, 88, 91
project plans, develop, 10
project staff, general development of the, 32
project strategies, develop, 3, 11
project teams, software, 23
Pyster, Arthur, 4, 5, 60

quality assurance, software, 27, 49, 50
quality metrics, 51 *(See also quantity metrics)*
quality-replacement matrix, 43

Reifer, Don J., ix
resolve conflicts, 3, 36, 41
responsibility, 26, 65
reviews, milestone, 47
Rue, Leslie W., 5
rules, 11, 15

Sacramento Bee, 7, 53
Sacramento State University, xi, 20
SEID Software Development Policy, 14
self-motivation, 30
set objectives, 3, 10
skills versus management levels, 76
software, overrun on the C-17, 53
software configuration management plans, 16
software crisis, 3, 6, 8, 53
Software Development Policy by TRW/SEID, 14
software development teams, build, 3, 39
software engineering economics, ii, 28, 57, 101
software engineering standards, 45
software engineering team, 39
software personnel, assimilate newly assigned, 31
software quality-assurance plans, 16
software system engineers, 27
staff organizations, 19
staff turnover, negative consequences of, 42
staff turnover, positive consequences of, 42
staffing, v, 2, 3, 5, 27, 28, 29, 30, 65, 73, 90, 102

Notes

NOTES

NOTES

www.ingramcontent.com/pod-product-compliance
Lightning Source LLC
Chambersburg PA
CBHW082103210326
41599CB00033B/6567